World Scientific Series in Current Energy Issues   Volume 6

Energy Efficiency

Innovations: Driving Prosperity, Slashing Emissions

World Scientific Series in Current Energy Issues

Series Editor: Gerard M Crawley *(University of South Carolina & Marcus Enterprise LLC, USA)*

Published

Vol. 6 *Energy Efficiency*
Innovations: Driving Prosperity, Slashing Emissions
edited by Henry Kelly

Vol. 5 *Critical Materials: Underlying Causes and Sustainable Mitigation Strategies*
edited by S Erik Offerman

Vol. 4 *Energy Storage*
edited by Gerard M Crawley

Vol. 3 *Energy from the Nucleus: The Science and Engineering of Fission and Fusion*
edited by Gerard M Crawley

Vol. 2 *Solar Energy*
edited by Gerard M Crawley

Vol. 1 *Fossil Fuels: Current Status and Future Directions*
edited by Gerard M Crawley

World Scientific Series in Current Energy Issues Volume 6

Energy Efficiency

Innovations: Driving Prosperity, Slashing Emissions

Editor

Henry Kelly

 **World Scientific**

NEW JERSEY · LONDON · SINGAPORE · BEIJING · SHANGHAI · HONG KONG · TAIPEI · CHENNAI · TOKYO

Published by

World Scientific Publishing Co. Pte. Ltd.

5 Toh Tuck Link, Singapore 596224

USA office: 27 Warren Street, Suite 401-402, Hackensack, NJ 07601

UK office: 57 Shelton Street, Covent Garden, London WC2H 9HE

Library of Congress Cataloging-in-Publication Data

Names: Kelly, Henry, 1945–　editor.

Title: Energy efficiency : innovations : driving prosperity, slashing emissions /
    editor, Henry Kelly (Boston University, USA).

Description: New Jersey : World Scientific, [2020] | Series: World scientific series in
    current energy issues, 2425-0171 ; volume 6 | Includes bibliographical references.

Identifiers: LCCN 2020010947 | ISBN 9789811217852 (hardcover) |
    ISBN 9789811217869 (ebook for institutions) | ISBN 9789811217876 (ebook for individuals)

Subjects: LCSH: Energy conservation--Technological innovations. |
    Energy consumption--Technological innovations.

Classification: LCC TJ163.3 .E22 2020 | DDC 621.042--dc23

LC record available at https://lccn.loc.gov/2020010947

British Library Cataloguing-in-Publication Data

A catalogue record for this book is available from the British Library.

For any available supplementary material, please visit
https://www.worldscientific.com/worldscibooks/10.1142/11761#t=suppl

Desk Editors: Balamurugan Rajendran/Amanda Yun

Typeset by Stallion Press
Email: enquiries@stallionpress.com

Preface

Strategies for encouraging innovation in energy efficiency lie at the heart of any strategy with a chance of meeting global goals in climate change. The ideas discussed in this volume demonstrate that opportunities to improve efficiency worldwide are enormous. We are not close to approaching theoretical limits.

Economics will, of course, limit what can be achieved, but there is reason to be optimistic. In current markets, it is almost always less expensive to invest in efficient energy use than new energy production. There are limits, of course, but a long history of low energy prices and energy production subsidies has left a legacy of buildings, transportation systems, and industry that are far below efficiency levels that make economic sense, particularly if the very real economic cost of climate change is taken into account. Unlike most new power production systems, efficiency systems can be made in manageable increments instead of the multi-billion-dollar investments typical of many traditional power plants. Modular investments typically facilitate continuous improvement in technology and systems design. And, while they are not completely free of risk (some new devices, for example, now rely on expensive metals), efficiency systems do not typically present significant environmental or safety risks. Efficiency investments made to meet climate goals can improve the quality of life for most people: buildings can be more comfortable with better lighting and cleaner air, and fast, reliable travel can reduce congestion and improve the air quality.

This volume is not designed to develop global scenarios of potential efficiency investments in the coming years — this has been well covered in other sources, particularly work explored in Chapter 7. What this will do is provide a strategic look at energy efficiency opportunities, many of which have not yet been incorporated into formal models. This is not an easy task since energy use touches virtually every part of the global economy. This volume will pursue three basic themes:

1. The need to consider *energy use as an integrated system*: The net efficiency of providing thermal comfort, lighting, and other amenities in buildings, for example, depends on the efficiency of the system's equipment, the way the building is designed (i.e., whether it allows daylighting), the way it is operated, and the design of the urban area where the system is located.

2. The shift to *electric systems*: Since it is not practical to capture carbon dioxide and other greenhouse gases from cars, buildings, and most manufacturing processes, new ways need to be found to power most of these operations. Electricity produced in low- or zero-emission systems to reduce greenhouse emissions is usually a preferred option.

3. The rapidly *growing role of information technology*: The collection and analysis of data is now a major part of energy systems and critical to achieving efficiency goals. This applies both to equipment (e.g., computers in cars) and integrated systems (e.g., algorithms that dispatch vehicles to maximize mobility).

The first chapter provides an overview of these topics and some background on the way energy is used today. The second chapter provides a strategic look at efficiency gains in buildings: building equipment, integrated building designs, and new control strategies — including strategies for integrating building controls with smart electric grids. The third chapter covers efficiency in industry, including novel (often electric-based) processes, improved product designs that can reduce material use and energy use when the product is used. This often involves use of novel materials, including bio-inspired materials. The fourth chapter focuses on new, highly efficient approaches to transportation. This includes dramatic new business models and vehicle systems that have the potential to revolutionize mobility (e.g., shared mobility, electric vehicles, and automated vehicles). The fifth chapter takes a close look at the potential for improving the efficiency of air travel and airfreight, a transportation sector that is growing rapidly and presents unique challenges for reducing emissions. The sixth chapter takes the concept of systems to an even broader level by exploring how urban areas use energy, and assesses the impact of urban design on integrated efficiency. The final chapter reviews the work that has been done to incorporate assumptions about efficiency into overall estimates of global energy use and greenhouse gas emissions.

Energy efficiency has already made an enormous contribution to meeting climate goals. But, as the analysis presented in this volume will demonstrate, we are far from running out of opportunities. Roadmaps to meet climate goals will all require dramatic new investments in efficiency, starting immediately. We hope that this volume will provide both hope that this can be achieved and the ideas that can guide this process.

Foreword to the World Scientific Series on Current Energy Issues

Sometime between four hundred thousand and a million years ago, an early humanoid species developed the mastery of fire and changed the course of our planet. But as recently as a few hundred years ago, the energy sources available to the human race remained surprisingly limited. In fact, until the early nineteenth century, the main energy sources for humanity were biomass (from crops and trees), their domesticated animals and their own efforts.

Even after many millennia, the average per capita energy use in 1830 only reached about 20 Gigajoules[a] (GJ) per year. By 2010, however, this number had increased dramatically to about 80 GJ per year.[1] One reason for this notable shift in energy use is that the number of possible energy sources increased substantially during this period, starting with coal in about the 1850s and then successively adding oil and natural gas. By the middle of the twentieth century, hydropower and nuclear fission were added to the mix. As we move into the twenty-first century, there has been a steady increase in other forms of energy such as wind and solar, although presently they represent a relatively small fraction of world energy use.

Despite the rise of a variety of energy sources, per capita energy use is not uniform around the world. There are enormous differences from country to country, pointing to a large disparity in wealth and opportunity (see Table 1). For example, in the United States the per capita energy use per year in 2017 was 301.2 million Btu[2] (MMBtu) and in Germany, 169.5 MMBtu. In China, however, per capita energy use was only 57.1 MMBtu in 2007, but grew dramatically to 138.7 MMBtu in 2017.

---

[a]GJ = 0.947 MMBtu.

Table 1.   Primary energy use per capita in million Btu
(MMBtu).[2]

Country	2007 (MMBtu)	2017 (MMBtu)	Percentage change (%)
Canada	416.1	411.7	−1.1
United States	336.9	301.2	−10.6
Brazil	52.7	60.4	+14.6
France	175.7	154.1	−12.3
Germany	167.8	169.5	+1.0
Russia	204.0	275.6	+35.1
Nigeria	6.1	8.1	+32.8
Egypt	36.4	41.6	+14.3
China	57.1	138.7	+142.9
India	17.0	22.7	+33.5
World	**72.2**	**77.3**	**+7.1**

India, also saw a significant increase in per capita energy use from 2007 to 2017 of 33.5%. The general trends over the last decade suggest that countries with developed economies generally show modest increases or even small decreases in energy use, but that many developing economies, particularly China and India, are experiencing rapidly increasing energy consumption per capita.

These changes, both in the kind of energy resource used and the growth of energy use in countries with developing economies, will have enormous effects in the near future, both economically and politically, as greater numbers of people compete for limited energy resources at a viable price. A growing demand for energy will have an impact on the distribution of other limited resources such as food and fresh water as well. All this leads to the conclusion that energy will be a pressing issue for the future of humanity.

All energy sources have disadvantages as well as advantages, risks as well as opportunities, both in the production of the resource and in its distribution and ultimate use. Coal, the oldest of the "new" energy sources, is still used extensively to produce electricity, despite its potential environmental and safety concerns in both underground and open cut mining. Burning coal releases sulfur and nitrogen oxides which in turn can lead to acid rain and a cascade of detrimental consequences. Coal production requires careful regulation and oversight to allow it to be used safely and without damaging the environment. Even a resource like wind energy, which uses large wind turbines, has its critics because of the potential for bird kill and noise pollution. Some critics also find large wind turbines an unsightly addition to the landscape, particularly when the wind farms are erected in pristine environments. Energy from nuclear fission, originally believed to be "too cheap to meter"[3] has not had the growth predicted because of the problem with long term storage of the waste from nuclear reactors and because of the public perception regarding the danger of catastrophic accidents such as happened at Chernobyl in 1986 and at Fukushima in 2011.

Even more recently, the amount of carbon dioxide in the atmosphere has steadily increased and is now greater than 400 parts per million (ppm).[4] The amount of another gas, methane (CH_4), in the atmosphere is also increasing and is an even more potent greenhouse gas than CO_2. Methane is often released as part of the extraction of oil. The increase of these greenhouse gases has raised concern in the scientific community about the continued use of fossil fuels and has led the majority of climate scientists to conclude[5] that this will result in a significant increase in global temperatures. We will see a rise in ocean temperature, acidity and sea level, all of which will have a profound impact on human life and ecosystems around the world. Relying primarily on fossil fuels into the future may therefore prove precarious, since burning coal, oil and natural gas will necessarily increase CO_2 levels. Certainly, for the long term, adopting a variety of alternative energy sources which do not produce CO_2 nor release methane, seems to be our best strategy.

In addition, we should consider ways to use energy more efficiently, including better insulation of our buildings, more energy efficient manufacturing, and much more energy efficient modes of transportation. As noted in Table 1, there remains as much as a factor of two in energy use per capita even among developed economies. As energy becomes more expensive in the future, this will undoubtedly provide additional incentives for more efficient energy use.

The volumes in the World Scientific Series on Current Energy Issues explore different energy resources and issues related to the use of energy. The volumes are intended to be comprehensive, accurate, current, and include an international perspective. The authors of the various chapters are experts in their respective fields and provide reliable information that can be useful to scientists and engineers, but also to policy makers and the general public interested in learning about the essential concepts related to energy. The volumes will deal with the technical aspects of energy questions but will also include relevant discussion about economic and policy matters. The goal of the series is not polemical but rather is intended to provide information that will allow the reader to reach conclusions based on sound, scientific data.

The role of energy in our future is critical and will become increasingly urgent as the world's population increases and the global demand for energy turns ever upwards. Questions such as which energy sources to develop, how to store energy and how to manage the environmental impact of energy use will take center stage in our future. The distribution and cost of energy will have powerful political and economic consequences and must also be addressed. How the world deals with these questions will make a crucial difference to the future of the earth and its inhabitants. Careful consideration of our energy use today will have lasting effects for tomorrow. We intend that the World Scientific Series on Current Energy Issues will make a valuable contribution to this discussion.

References

1. Our Finite World: World energy consumption since 1820 in charts. March 2012. Accessed in February 2015 at http://ourfiniteworld.com/2012/03/12/world-energy-consumption-since-1820-in-charts/.

2. U.S. Energy Information Administration, Independent Statistics & Analysis. Accessed in Feb 2020 at http://www.eia.gov/cfapps/ipdbproject/iedindex3.cfm?tid=44&pid=45&aid=2&cid=regions&syid=2005&eyid=2011&unit=MBTUPP.

3. www.nrc.gov/docs/ML1613/ML1613A120.pdf. Remarks prepared by Lewis L. Strauss, Chairman, US Atomic Energy Commission, 16th Sep, 1954. Accessed in Feb 2020. There is some debate as to whether Strauss actually meant energy from nuclear fission or not.

4. NOAA Earth System Research Laboratory, Trends in Atmospheric Carbon Dioxide. Accessed in March 2015 at http://www.esrl.noaa.gov/gmd/ccgg/trends/.

5. IPCC, Intergovernmental Panel on Climate Change, Fifth Assessment report 2014. Accessed in March 2015 at http://www.ipcc.ch/.

Contents

Contents

https://doi.org/10.1142/9789811217869_0001

Chapter 1

Energy Efficiency: An Introduction

Henry Kelly

*Institute for Sustainable Energy, Boston University,
Boston, MA 02215, USA*
henry.c.kelly@gmail.com

1. Introduction

Energy use is an integral part of most economic systems, and energy efficiency technologies are intimately connected to the innovations needed to build competitive, dynamic national economies and create rewarding employment opportunities. Policies affecting energy efficiency or energy productivity are, as a result, an essential element of any strategy for achieving global, national and local economic, energy and environmental goals. Systems optimized to produce what people value — comfortable interior spaces or fast, reliable transportation — include strategic investments in both energy supplies and the systems that use energy to meet our needs. The efficiency of most energy systems operating today, however, is much lower than levels that would be justified economically. They are often well below levels that would be optimum for society as a whole, and are far below theoretical limits. This volume will explore the technical opportunities to improve system efficiency, the implications of optimum levels of investments in efficiency, and the steps that can be taken to encourage wise investments.

Successful firms constantly look for ways to minimize production costs, including the cost of energy. This raises an obvious question: why focus special attention on the efficiency of energy use as opposed to the consumption of any other factor of production? The primary reason is that the price paid for energy falls far short of the actual cost of energy use.

Urban air pollution, climate change and other environmental costs, and the hidden cost of international security are not included in the prices consumers and businesses pay for energy. This is a major failing in today's society since fossil energy production and use are responsible for a large fraction of the emissions contributing to urban air pollution and global warming, and the social costs created by these emissions.

The underpricing of energy can only be rectified by public policy. The most obvious solution, and the one universally supported by economists, is to simply raise the price of energy to a level where environmental and other costs are included in the price and therefore in the decisions made by businesses. The political difficulty of doing this is obvious and it is also not clear that any politically feasible price change will be adequate to make the rapid changes in energy use needed to meet the goals set in Paris for future greenhouse emissions. Ongoing decisions to keep energy inexpensive has led to deep distortions in economies around the world. These include things as fundamental as where people choose to live and travel — choices that cannot be changed quickly. Ironically, attractive investments in energy efficiency are often ignored because energy consumption cost is often a small fraction of the overall cost of operating a business or household (US household energy expenditures make up ~5% of disposable income).[a] Poor decisions may also reflect a simple lack of information. Before the government-mandated energy consumption data labels, for example, consumers had no convenient way to compare the efficiency of refrigerators or automobiles.

Despite the limitations, investments in energy efficiency have had spectacular results. The International Energy Agency estimates that energy efficiency improvements that occurred between 2000 and 2017 reduced 2017 energy demand by 12%. Their "Efficient World Scenario" suggests that adoption of "all cost-effective energy efficiency opportunities" would allow the global economy to double with only a "marginal increase in primary energy demand".[b] Increases in energy efficiency do not necessarily translate into reductions in energy and emissions, since demand for energy increases with income levels. The savings that result from efficiency gains lower the price of services and further contribute to growing demand.

---

[a]EIA, "Consumer energy expenditures are roughly 5% of disposable income, below long-term average", Energy Information Administration, 2014. https://www.eia.gov/todayinenergy/detail.php?id=18471#.

[b]IEA, 2018, Energy Efficiency: Analysis and Outlooks to 2040, https://webstore.iea.org/download/direct/2369?fileName=Market_Report_Series_Energy_Efficiency_2018.pdf.

Developing countries present a particular challenge as increased prosperity drives increases in amenities such as automobile ownership and air conditioning. Demand growth has outstripped gains in energy efficiency, resulting in increasing energy use. In wealthy countries, efficiency gains have kept pace with growing demand. As a result, total energy use has been relatively constant for over a decade. However, global efficiency improvements are not keeping pace with economic growth. Global energy use and emissions increased in 2017.[c]

The decrease in the energy needed to operate the economy ("energy intensity") results from two primary factors[1]: innovative technologies that improve efficiency and designs, and changes in the structure of the economy from businesses that use large amounts of energy (e.g., iron and steel) to businesses with lower energy needs (e.g., banking and insurance). In some nations imports and exports can also play a major role, for example, US imports of energy-intensive products like steel reduced domestic energy use. Structural change in an economy is typically considered to be independent of energy efficiency, yet the boundary is increasingly unclear. One clear trend, for example, is "demassification" of modern economies, which can be driven by smaller office and living spaces, internet shopping, and the ability of improved designs and new materials to provide improved services for a given weight of materials. A growing fraction of the value of an electric car, for example, results from computers, sensors and software designed to control the operation of the vehicle, increase safety, minimize use of materials and support passenger comfort. These innovations can also dramatically reduce the energy needed to provide mobility services. New urban designs, driven largely by forces independent of energy costs, can also lead to dramatic reductions in transportation energy use.

This chapter examines both efficiency technologies and the structural changes that can drive changes in the net energy intensity of the economy. After a brief review of the basic statistics of energy use and the history of efforts to improve energy efficiency, it will explore three primary themes: the operation of integrated energy systems (efficient equipment operating in optimized networks to deliver services like mobility), electrification (driven

---

[c]IEA, "Energy efficiency 2018", International Energy Agency, 2018a. https://webstore.iea.org/download/direct/2369?fileName=Market_Report_Series_Energy_Efficiency_2018.pdf.

in part by environmental concerns), and digitization (information technologies, communications, data and optimization). While global data will be used where available, greater detail is typically available for the US and other affluent economies, and our discussion necessarily reflects this asymmetry.

2.   Some Basics

Figure 1 shows how energy is used worldwide and in the relatively affluent Organization for Economic Co-operation and Development (OECD) countries. The OECD was responsible for 42% of global energy use in 2015 and estimates suggest that this will shrink to 30% by 2050. It also shows the US Energy Information Administration's "reference case" estimates of growth (many other estimates will be discussed in later chapters but these traditional estimates show the scope of the problem and can help set priorities).

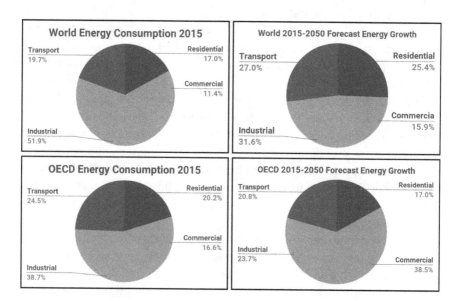

Fig. 1.   Energy use and growth in energy use.

Source: EIA 2017 International Energy Outlook Reference Case. World energy in 2015 = 611 EJ, 579Q, World 2015–2050 growth = 246EJ, 233 Q. OECD 2015 energy = 255EJ, OECD 2015–2050 growth = 255EJ, 242Q. Industrial energy use includes energy used to heat and air-condition and provide lighting for industrial buildings. HVAC is 15% of industrial energy use in Europe (Rehfeldt, 2018).

The first thing to notice is that growth, following this estimate, will see the world's energy needs growing 83% by 2050. This would make it all but impossible to meet our environmental goals. Enormous investments in clean energy sources and investments designed to capture and store carbon dioxide would be needed.

Over half of today's energy use goes towards industry, much of it in inefficient facilities outside the OECD, particularly in China and India. The OECD's energy use is dominated by transportation and buildings. More than half of the worldwide energy consumption involves production of iron, steel and cement.[d] Much of this is being used to build infrastructure in developing nations. The world's production of steel doubled between 2000 and 2017, and three quarters of this growth was in China.[e]

World energy growth patterns suggest that most of the nations not in the OECD are moving in a direction that would leave them with similar energy use patterns as that of OECD's today. Demand for transportation energy grew 45% between 2000 and 2017, driven by global wealth increases.[f]

Retirements of obsolete equipment and a shift in global economic structure led to estimates showing 70% of global energy growth, and 75% of the OECD's growth results from growth in energy use for transportation and residential and commercial buildings (including the business service operations delivered from these buildings).

Roughly half of the electricity used worldwide is in industry, with most of the rest used in buildings (Fig. 2). In the OECD, buildings use about two-third of all electricity (buildings make up 78% of the electricity demand in the US; this includes electricity used to supply lighting, heating and cooling to industrial buildings). Buildings are responsible for consuming 28% of the world's energy, but this figure accounts for over 40% of the energy in the OECD. This pattern could change dramatically as non-OECD nations modernize and if climate policies lead to rapid electrification in all sectors. Developing regions acquire more modern appliances, particularly air conditioning. Demand for air conditioning has doubled since 2000 and growth is likely to continue given increasing global

---

[d]IEA, "Tracking clean energy progress 2017", International Energy Agency, 2017. https://www.iea.org/publications/freepublications/publication/TrackingCleanEnergy Progress2017.pdf.
[e]IEA, "World energy investment", International Energy Agency, 2018b. https://webstore.iea.org/download/direct/1242?fileName=WEI2018.pdf.
[f]IEA, 2018a.

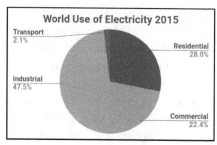
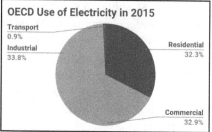

Fig. 2.   Electricity use.

Source: EIA 2017 International Energy Outlook Reference Case. World energy in 2015 = 611 EJ, 579Q, World 2015–2050 growth = 246EJ, 233 Q. OECD 2015 energy = 255EJ, OECD 2015–2050 growth = 255EJ, 242Q. Industrial energy use includes energy used to heat and air-condition and provide lighting for industrial buildings. HVAC is 15% of industrial energy use in Europe (Rehfeldt, 2018).

temperatures, and that only 8% of the people living in the globe's hottest regions have access to air conditioning.[g] Demand for transportation energy grew 45% between 2000 and 2017, driven by global wealth increases.[h]

3.   A Brief History

The pursuit of efficient use of inputs is an inherent part of any economy, but the industrial revolution of the 19th century introduced technologies that drove huge growth in energy use and options for using it efficiently. One of the heroes of this era was James Watt. He did not invent the steam engine, though he made it practical by increasing the efficiency of steam power by a factor of three. Watt not only made a huge improvement in efficiency, he also invented a business model that has resurfaced in recent years. He did not sell steam engines; he sold energy savings instead. His fee was a percentage of the saved fuel costs.[2] Calculating these savings required careful measurement of energy flows, and his creativity in doing this led to the system of units and measurements, which was memorialized by its basic unit of power (Watt) being named after him.

Watt's inventions were, of course, only the beginning of efficiency improvements since these were far lower than the theoretical maximum

---

[g]IEA, "The future of cooling", International Energy Agency, 2018c.
[h]IEA, 2018a.

efficiency. His 1778 engine was about 2.5% efficient.[3] Engines reached 9.6% efficiency by 1826 and steam turbines are now more than 40% efficient.

Government interest in energy policy largely centered on taxing production and securing energy supplies during the two World Wars. After the Second World War energy became relatively inexpensive and the promise of "meterless" nuclear electricity made considerations of efficiency seem irrelevant. Buyers seldom knew anything about the energy consumption of the equipment they purchased.

In the US, post WWII energy research was supported by the Atomic Energy Commission, a direct outgrowth of the US nuclear weapons program that had little interest in efficiency. The oil embargos of 1967 and 1973 changed this overnight. New institutions were created in governments worldwide. In the US, energy policy was taken out of the Atomic Energy Commission and given to a new office with a much broader scope.[i]

Perhaps the single most influential analysis of energy policy in this era was the Ford Foundation's 1974 Energy Policy Project directed by S. David Freeman,[4] which broke new ground in many areas, including a focus on energy efficiency (then called "energy conservation"). It emphasized that "... using energy more efficiently so that a slowdown in this sector will not seriously impair economic growth and job opportunities .... In this book we hope to demonstrate that slower energy growth than [what] we have recently experienced can work without undermining our standard of living, and can also exert a powerful positive influence on environmental and other problems closely intertwined with energy ... a long term slowdown in the energy growth rate actually has the effect of increasing employment". The discussion of environmental benefits included the observation that "the burning of fossil fuels may lead in the near future to global climatic change".

The same report outlined a roadmap for public policy in energy efficiency that is still relevant today. It includes discussions of energy taxes, fuel economy standards, building codes, the "expansion of urban mass transit systems and development of a system of bikeways", and high-speed rails. It is remarkable that its most ambitious energy forecast (its "zero energy growth" scenario) almost predicted the actual trajectory of US energy consumption. Recognizing that this scenario depended

---

[i]US DOE, "Timeline of events: 1951–1970", US Department of Energy. https://www.energy.gov/lm/doe-history/doe-history-timeline/timeline-events-1951-1970.

heavily on innovation, the report laments that "in the past, government R&D funding has been confined to new sources of supply, primarily atomic energy. Almost nothing has gone into energy conservation", and it argued for a new set of research priorities. Many of the recommendations of the report were translated directly into the National Energy Conservation Policy Act,[j] which established fuel economy standards and many of the other foundations of energy efficiency policy.

Obviously, there was much new work to be done. A few early studies set a clear direction. A group of physicists[k] tackled the challenging question of how energy was actually used in the economy and how close contemporary processes (e.g., lighting, heating buildings and travel) came to the maximum theoretical ability of the energy source to do work. The results were breathtaking. Energy costs often played almost no role in the design of products like refrigerators or cars and they operated at a small fraction of what was theoretically possible. Many of the authors, including Berkeley's Arthur Rosenfeld, abruptly changed their careers and helped forge new research organizations built around themes like energy efficient buildings. A widely read article "Energy Strategy: The Road Not Taken" in *Foreign Affairs* by Amory Lovins made the arguments accessible and compelling.[l]

What followed was dramatic. While the US's energy intensity had been slowly increasing since 1949, dramatic increases began by the mid-1950s (Fig. 3). Much of this, of course, resulted from the sharp increase in energy prices driven by the oil embargo, together with new patterns of trade economic structure, but policy, and a new awareness of the economic benefits of energy efficiency, ensured that efficiency improvements continue while prices fluctuate.[m]

But much remains to be done. Energy efficiency, which is still struggling to find a share in national energy research budgets, barely exceeded 20% in 2017[n] (Fig. 4).

---

[j] https://www.govinfo.gov/content/pkg/STATUTE-92/pdf/STATUTE-92-Pg3206.pdf#page=83.

[k] 1974 American Physical Society summer study entitled *Efficient Use of Energy: A Physics Perspective*. Many of the authors reconvened to prepare the 1981 report "A new prosperity, building a sustainable energy future: the seri solar conservation study".

[l] A.M. Lovins, "Energy strategy: the road not taken", *Foreign Affairs*, October 1976. https://www.foreignaffairs.com/articles/united-states/1976-10-01/energy-strategy-road-not-taken.

[m] EIA, "Monthly energy review", Energy Information Administration, 2018. https://www.eia.gov/totalenergy/data/monthly/pdf/sec1_16.pdf.

[n] IEA, 2018b.

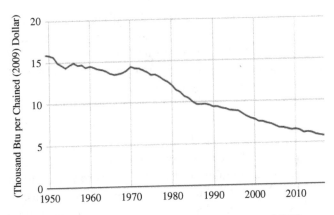

Fig. 3.   Total US primary energy consumption per real dollar of GDP.
Source: EIA (2008).

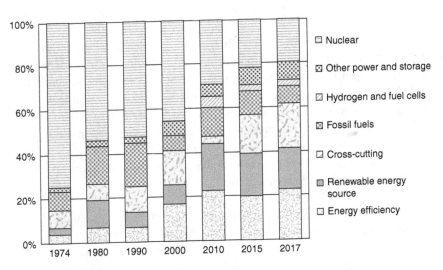

Fig. 4.   Energy technology R&D budgets.
Source: IEA, Energy Technology's 2018 R&D budgets.

4.   Major Themes

The core challenge of energy efficiency is to deliver amenities like comfortable interior spaces and rapid, safe transportation at the lowest cost (including environmental and other non-market costs). These sectors will

be explored in detail in later chapters. The discussion here will focus on
three themes that are threaded throughout: design and operation of integrated systems, the critical role of information technology, and electrification. These themes have gone from the periphery to the center of analyses
exploring the potential of energy efficient economies

4.1. *Systems*

The energy intensity of an economy results from the operation of integrated systems, which combine to deliver services to final consumers. The
systems include individual technologies, design and operations. In the case
of buildings, systems also involve integration with natural systems — environmental sources of light, heat, ventilation, energy reservoirs in the
ground under the buildings, and shade and evaporative cooling from growing plants.

In principle markets should be able to optimize investments, both in
components and entire systems. In practice, however, optimization is limited by a maze of non-market factors — particularly where systems
depend on both private and public investments. Examples may include the
design of highway systems or regulations that prevent electric generating
utilities from investing on the customer's side of electric meters. The
design and operation of these systems are very different in different parts
of the economy. Later discussions will explore these systems in greater
detail but a few highlights should set the stage.

The system supporting comfortable living and working spaces in
buildings depends on investments in:

- *Efficient components* (insulating materials, windows, heating and cooling equipment, lighting fixtures and efficient appliances),
- *Building designs* that lend themselves to efficient operation and that
  skillfully integrate buildings with their environment. This integration
  can include orienting and operating windows to ensure optimum flows
  of light and heat, designs for getting daylight deep into interior spaces,
  optimum use of natural circulation, selecting the size of windows facing
  different directions, and many others.

  Sensor and control systems adjusting heating and cooling, ventilation, lighting, and other building services to meet occupant needs while
  minimizing energy use. These control systems can also ensure that
  buildings are operating as designed and detect anomalies during

operations so that equipment failures and other problems that can impact efficiency (and comfort) are quickly located and repaired.

In addition to these site-based systems, buildings are tightly coupled with urban, regional and global systems.

- *Urban environmental systems* can be strongly influenced by building operations, including a building's access to ventilation and natural lighting. Climate change, driven in part by building energy use, will increase average temperature levels in many regions that see increasing use of air conditioning. Although climate change goals embodied in the 2016 Paris agreement[o] include keeping average increases below 1.5–2°C, temperature increases could be much higher if policies fail. Large regions may be uninhabitable without air conditioning. Urban areas present a particular challenge since heat-absorbing roads and roofs and other factors push city temperatures well above that of the surrounding countryside.[p] Increased use of air conditioning will contribute to the problem as these systems pump heat out of buildings and into the environment. Air conditioning alone can increase the temperature of cities by as much as 1°C.[q] Increased outdoor temperatures increase the demand for cooling as well as decrease the efficiency of cooling units. This creates a vicious cycle. Many of these self-reinforcing negative feedback effects can, of course, be reversed and become positive reinforcing effects given effective policies.
- *Electric grid systems* are tightly linked to buildings since buildings are a major part of total electric demand (78% in the case of the US). Managing electric peak loads and making effective use of intermittent renewable electricity, such as wind and solar, will require a way to integrate grid controls with building controls. Building mass can provide significant amounts of very *low-cost energy storage* when integrated into a smart grid management system. They also provide opportunities for installing ice-storage and other thermal storage that cost far less than traditional "electric storage" devices.[5] The need for

---

[o]United Nations Climate Change, The Paris Agreement, 2018. https://unfccc.int/process-and-meetings/the-paris-agreement/the-paris-agreement.

[p]EPA, "Reducing urban heat islands: Compendium of strategies", Environmental Protection Agency, 2012. https://www.epa.gov/heat-islands/heat-island-compendium.

[q]IEA, 2018c. https://webstore.iea.org/the-future-of-cooling.

building-grid coordination will increase as climate policies replace equipment that use fossil fuels with electric alternatives such as electric heat pumps.

Transportation systems have for generations depended on systems of publicly operated highways and privately-owned vehicles. In most regions, public transportation, rail, and air travel play a comparatively small role in overall energy use in passenger transportation. In International Energy Agency (IEA) member countries, automobiles are responsible for nearly 60% of all energy use. But these mobility systems are being challenged by a range of innovations based on shared mobility models, electrification, and potentially vehicle automation. These new systems have the potential to cut the cost of trips, increase their speed, and make transportation available to people who are not well served by existing transportation systems for reasons of income or disability. They could also present a challenge for energy intensity: the efficient new systems could cut the cost of mobility, thereby increasing demand for travel offsetting the efficiency gains and leading to an overall increase in transportation energy use.

The new mobility systems include:

- *Vehicle equipment efficiencies* will increase as designs make increasing use of lightweight materials, aerodynamic design, low rolling resistance, and efficient motors and controls that optimize vehicle performance (e.g., controlled acceleration and braking). Sensors and automated controls ensure optimum performance of engines and other vehicle systems. Taken together, these improvements can increase the fuel efficiency of fuel-powered vehicles by factors of 2–3,[r] but while these improvements can provide near-term reductions in emissions, meeting 2050 climate goals will require a massive shift towards electric vehicles.

- Private vehicle owners increasingly choose to drive alone and park their vehicles for extended periods of time. This has led to inefficient use of roads. Congestion greatly increases travel times and result in poor utilization of vehicles (most cars are stationary more than 95% of the time). It is now possible to imagine a system that provides faster, more reliable, safer and less expensive mobility using available technologies

---

[r] US DOE, "Quadrennial technology review 2015", US Department of Energy, 2015. https://www.energy.gov/quadrennial-technology-review-2015.

that include mobile phones with GIS position locators and high-speed data links. These can be coupled with sophisticated dispatch systems based on new data management systems and advances in the mathematics of route optimization. Routing information can be sent to vehicles and highway control systems (e.g., traffic lights, speed limits and tolls).

Recognizing the dramatic changes that may reshape personal transportation over the next few decades, most major automobile manufacturers are making major investments in business models that do not depend on selling cars but on selling "mobility as a service" (MAAS). This means that they will sell customers a safe, fast and convenient trip to their destination. Depending on an analysis of regional demand, this could either mean an on-demand individual pickup and drop-off (e.g., Uber, Lyft or a specialized fleet of vans), a just-in-time rental of a scooter, or a bicycle trip. Vehicle use would be optimized by using on-demand systems to connect with hubs of high-speed buses or trains. Customers would have options with different levels of speed, convenience and price. Public transportation companies are considering similar offerings.[6,s] Models of such systems show the potential to greatly reduce congestion, increase speed trips and cut costs. An examination of taxi trips in New York city, for example, showed that a rationalized dispatch system could have moved the same number of people to their destinations at the same speed with one-seventh the number of vehicles on the road.[7]

- Automated, connected vehicles could dramatically change the economics of MAAS by eliminating the cost of drivers (see Fig. 5). Automation (used here to mean vehicles with no drivers) could eliminate the need for parking, reduce the number of accidents, and provide mobility to people unable to drive for reasons of age or disability. Connectivity implies the ability of vehicles to communicate with each other and with highways increasing safety, this opens options for using traffic signals to prioritize multi-passenger vehicles, change pricing to reduce congestion (or to give discounts to multi-passenger vehicles during periods of environmental emergencies), and to facilitate evacuations in case of disasters.

---

[s]UC Davis, Institute for Transportation Studies, "Pooling and pricing: harnessing the 3 revolutions to solve congestion, climate change, and social equity", Davis, California. https://3rev.ucdavis.edu/events/pooling-and-pricing-harnessing-3-revolutions-solve-congestion-climate-change-and-social.

Energy Efficiency

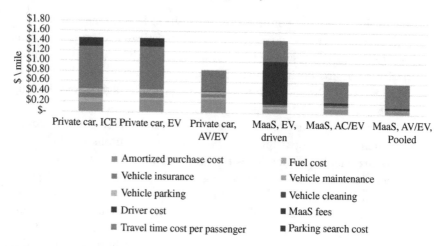

Fig. 5. Future trip costs.
Source: Fulton.[9]

A number of factors influence the net costs of automated vehicle trips and there are many unknowns. The greatest of these is uncertainty about how consumers will react to the dramatically new mobility services being offered. Will they accept rides in multi-passenger vehicles or vehicles without drivers? The estimates in Fig. 5[8] suggest that the cost of MAAS will be almost identical to conventional vehicle ownership. Will the conveniences offered be enough to drive new markets? With automation, costs are much lower but the use of multi-passenger vehicles depends critically on whether the perceived cost of greater travel time in multi-passenger vehicles exceeds the cost of having a single passenger pay for all vehicle costs. Although the analysis shown in Fig. 5 suggests that the cost of shared (pooled) and unshared rides are almost identical, many factors influence this outcome and there is considerable uncertainty about most of them.

The question relevant to energy efficiency, of course, is whether these system-wide changes in mobility will increase energy efficiency in highway transportation. Wadud *et al.*[9] Most assessments suggest that automation will increase efficiency by encouraging platooning, automatically achieving "eco-driving", improved crash avoidance, vehicle right sizing and shared mobility.[t] There is significant disagreement over whether automation could offset these gains by lowering driving costs and increasing access to transportation for people unable to use existing transportation systems for

---

[t]Wadud *et al.*[9]

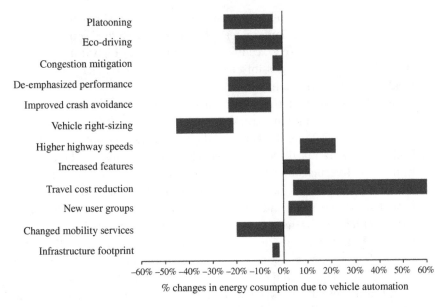

Fig. 6. Impact of vehicle automation on energy consumption.
Source: Wadud *et al.*[9]

reasons of disability, income, or other factors. One estimate is shown in Fig. 6. Recent analyses conclude that the energy efficiency of transportation systems can increase dramatically, but their total energy use could increase because dramatically reduced costs could drive large increases in travel (i.e., sleep while you commute long distances).[u] Large increases in convenience (particularly for the elderly, disabled and other underserved populations) will also drive demand. The power consumption and weight of automated vehicle systems could reduce system efficiencies by 2–4%, though it is hoped that new generations of equipment will be lighter and much more efficient.[10]

- *Trucks* are responsible for about a third of all transportation energy (5% of total global energy) and truck fuel use is growing rapidly (3% per year since 2000).[v] It is likely that this demand growth will continue, particularly in developing countries, but several technology-driven

---

[u]EIA, "Autonomous vehicles: uncertainties and energy implications", International Energy Agency, 2018. https://www.eia.gov/outlooks/aeo/pdf/AV.pdf.
[v]IEA, "The future of trucks", International Energy Agency, 2018d. https://www.iea.org/reports/the-future-of-trucks.

changes may soon disrupt the trend. These include shifts to a service economy, demassification, increased market share for sophisticated, lightweight products (e.g., cellphones), shipment of digital information instead of paper, and possibly use of 3-D printing that could increase local production of products.

Seventy per cent of the trucks on the road are light-duty vehicles, but heavy-duty, long-haul trucks are responsible for nearly two-thirds of all fuel used.[w] Long-distance freight movement has been revolutionized by using standardized containers that can be transferred efficiently from ships to trucks and rail. These container transport systems were one of the key drivers in the explosion of international trade over the past two decades. Standardized containers undoubtedly increased energy efficiency by moving more freight on highly efficient shipping platforms, and reduced energy use during loading and unloading processes. These savings were, however, overwhelmed by the massive increase in global shipping resulting from the cost savings resulting from the newly efficient system. In 1990, 35 million 40-foot equivalent containers were shipped. By 2014, 315 million were being shipped, some in leviathans capable of carrying 10,000 40-foot containers.[11]

Many strategies for moving freight through the last mile are being explored. Online shopping and 1–2 day delivery promises are driven by innovations in warehousing and retailing that have, in turn, decreased trips to physical stores and increased demand for delivery vehicles. Worldwide parcel delivery in 13 of the largest world economies grew 17% between 2016 and 2017. Each person in these countries now receives 22 parcels per year.[12] With so much uncertainty about demand and freight business models, the future of freight system energy efficiency remains unclear.

The use of GPS and rational routing algorithms by firms such as UPS, FedEx, DHL and Amazon have achieved major system-level improvements, but many inefficiencies remain. Recent estimates suggest that 10–20% of the energy used by light and medium weight trucks could be reduced with more efficient dispatching that will increase load factors and eliminate redundant trips (not including elimination of individual trips to stores).[x] This system-level opportunity has led to a range of innovations such as logistic service companies, and crowdsourcing

---

[w]IEA, 2018d.
[x]IEA, 2018d.

and collaboration platforms that have been introduced. Some innovations include the use of bicycles and small electric tricycles. Cities have also proposed limiting freight shipments to evening hours, which would reduce congestion and increase energy efficiency since freight vehicles could now spend less time on the road. Experiments are now underway, both by startups and established firms like Amazon, DHL and Uber.[y] Some business models include providing movement of both people and freight in urban areas. Attempts to combine personal and freight transportation has proven challenging since moving freight is much more complex and diverse than moving people. It is again difficult to understand the net effect on energy intensity; the pace of change has outstripped the ability of even the best analysts.

- *Aircraft* are responsible for a growing share of transportation energy consumption, and aviation energy use is growing rapidly. With the absence of policy intervention, jet fuel consumption could double by 2040.[z] In addition to the efficiency of the aircraft themselves, energy use in air travel is influenced by load factors (i.e., the fraction of the available seats filled), travel to and from airports, routing efficiency, and energy use on the ground. Absent increases in aircraft efficiency and better use of capacity that occurred between 2000 and 2017 , aircraft fuel use would be 68% higher than what it is today; improved flight routing could save an additional 5–10%.[aa] Some of the 6% of the aviation energy used on the ground (e.g., taxi-ing and waiting for clearance) could be saved by shifting to electric propulsion.

A recent study estimated that high speed rail could be competitive in trip time with 18% of commercial passenger trips (e.g., traveling to airports and security delays add up to approximately 210 minutes for an air trip and 80 minutes for a rail trip).[ab] Measured in terms of passenger miles per unit of energy, a shift to rail would lead to significant reductions in energy use per passenger mile. Aircraft fuel efficiency is expected to increase significantly in the coming decades but is unlikely to match the efficiency of rail travel.

---

[y] IEA, 2018d.
[z] EIA, "International energy outlook", Energy Information Administration, 2017. https://www.eia.gov/outlooks/ieo/pdf/0484(2017).pdf.
[aa] IEA, 2018a.
[ab] *Ibid.*

4.1.1 *Industry*

The diversity of industrial production systems makes any discussion of the energy efficiency of these systems difficult, but some examples can be useful:

- Additive manufacturing (or "3-D printing") affects energy use in several ways.[13] First, computer simulations can create designs that minimize the materials required to perform a function by using them to resist stress or strain only where they are needed. The additive manufacturing systems can produce complex shapes that would otherwise have been impossible to make via casting, machining or other conventional methods. Material savings can be as high as 70%, which translates into an equivalent saving in the energy embodied in the materials used in manufacturing. The low-mass products weigh less — a feature that can increase the energy efficiency of aircraft and other vehicles. The energy actually consumed by additive manufacturing is also lower than energy used by traditional methods. If modular manufacturing could reduce the economies of scale needed by traditional methods, it might be possible to produce devices in plants that can be located near consumers, thereby reducing the energy needed to transport warehouse components.

- Efficient production systems based on innovative information and computational systems are one of the core investments of advanced manufacturing. The US Department of Defense-supported Digital Manufacturing and Design Innovation Institute highlights four key elements of these systems:

    a. **Design, Product Development, and Systems Engineering:** Creating improved design tools and processes, integrating data across the manufacturing lifecycle, and developing automated manufacturing planning.

    b. **Future Factory:** Enabling digital integration and control in the manufacturing environment, and implementing tools to increase flexibility throughout the production cycle.

    c. **Agile, Resilient Supply Chain:** Facilitating access to digital information, supply chain visibility and design collaboration.

    d. **Cybersecurity in Manufacturing:** Designing and deploying assessment tools, and establishing a collaborative network for sharing best practices.[ac]

[ac] UI Labs Focus Areas. https://www.som.com/projects/ui_labs__digital_manufacturing_and_design_innovation_institute_dmdii.

- Production systems have increasingly become international with value being added to products in many different locations, thus making it increasingly difficult to track net efficiency of these linked systems. This is particularly true in electronics. Table 1[14] shows that no nation is responsible for more than 30% of the value of a cell phone. These international production networks result in part because the enormous reductions in global freight costs driven by box containers and other factors.

4.2. *Digitization*

Sensors, controls and the automated systems that link them have become major energy efficiency technologies in all parts of the economy. The power of these systems results both from innovations in equipment and in the mathematics behind the software that operates these systems. The IT tools optimize the performance of designs (minimizing material use and production processes), the performance of individual devices (e.g., vehicle engines), and the operation of entire systems, such as shared mobility networks and smart buildings.

Figure 7[ad] shows that investments in ICT (Information, Communications and Technology) grew rapidly in 2016 and 2017, and are now nearly two-thirds of all corporate investment in energy efficiency. Over 90% of the companies receiving these investments provide services for transportation or buildings. It is increasingly difficult to disentangle ICT investments from other efficiency investments since many of the efficiency technologies include information technologies.

Table 1.   Value added to iPhone 7.

US	$8.69
Japan	$67.70
Taiwan	$47.84
Unidentified	$21.80
Korea	$16.40
China	$8.46
Europe	$6.56

Source: Dedrick *et al.*[15]

[ad] IEA, 2018b.

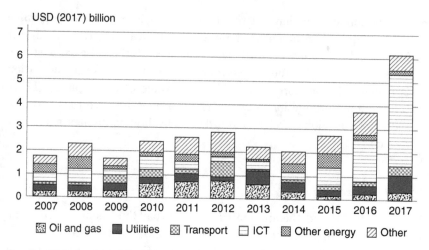

Fig. 7.   Global value of deals by corporate investors in energy efficiency technology. *Source*: IEA.

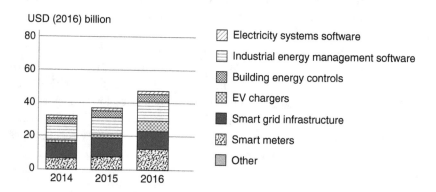

Fig. 8.   Investments in digital electricity infrastructure and software. *Source*: IEA, 2017.

Figure 8[ae] shows that many ICT innovations are moving from R&D concepts to global markets. Digital infrastructure investments made in global electricity infrastructure in recent years highlight the large investments being made in buildings and increasingly in charging electric

---

[ae]IEA, "Digitization & energy", 2017. https://www.iea.org/reports/digitalisation-and-energy. Accessed: 10 August 2020.

vehicles. These investments have become a major part of all global electricity investments; for comparison, global investment in gas-powered electric generation was about \$35 billion. This does not include IT investments in vehicles or MAAS systems. Automotive electronics, which were about 4% of the overall automobile production costs in 1970, are now about a third of production costs, possibly heading for 50% by 2030.[af] Automated, connected vehicles will greatly increase the value of IT equipment.

The energy consumed by information technologies themselves, however, may become a problem. Growth rates of information processing have been breathtaking. For example, the number of bytes transmitted increased by a factor of nearly 20 thousand between 1997 and 2017.[ag] This growth rate shows no sign of slowing and is already reflecting the emerging "internet of things", in which increasing numbers of devices are communicating with other devices. By 2015 there were over 600 million connected[15] devices. Some estimates suggest that there could be 27 billion connected devices by 2024.[15]

Taken together, the impact on electric demand could be significant — one estimate suggests that these information-related demands could exceed 20% of the worldwide electric demand by 2030. But there are reasons to be skeptical about such estimates. The energy used by ICT and entertainment and media enterprises has not increased over the past decade despite exponentially growing demand. There are two primary reasons for this:

(1) Huge increases in the energy efficiency of computing and the energy used to cool this equipment. Computations per kilowatt hour (kwh) doubled every 1.57 years between 1946 and 2009, and there is room for further improvements (one estimate suggested that the theoretical efficiency of computation is roughly a million times higher than present levels).[16]

(2) Improvements made to the air-conditioning systems used to cool computer facilities result in increased energy efficiency — these systems once required more energy than the computers themselves. Drawing

---

[af] "Statistica 2018". https://www.statista.com/statistics/277931/automotive-electronics-cost-as-a-share-of-total-car-cost-worldwide/.

[ag] https://www.cisco.com/c/en/us/solutions/executive-perspectives/annual-internet-report/index.html, https://www.cisco.com/c/en/us/solutions/collateral/executive-perspectives/annual-internet-report/white-paper-c11-741490.html.

on many of the innovations of building technologies, these demands have been dramatically reduced. Some large firms have also chosen to manufacture computers in cooler climates. Instead of doubling the electricity demand of a server facility, as was the case a few years ago, Google, for example, now claims that cooling adds only 12% to its overall computer electricity demand.[ah]

A second effect shaping the digitization of energy efficiency is structural. This has resulted in the shift from systems like PCs and large television screens to tablets and hand-held devices. Device-to-device systems, of course, require no screens and often do not need indicator lights either. It is also likely that extra energy savings are achieved by shifting from distributed local computing to energy-efficient server facilities. This has led to a significant decline in the overall electric demand in the IT sector. This result is one of many that makes clear definitions of efficiency difficult. If users prefer to receive information on small portable devices using cloud computing (in highly efficient computer centers) instead of stationary PCs and large televisions, is the resulting decline in energy use a gain in efficiency? Clearly the technology that made small devices increasingly powerful and easy to use played a decisive role.

At issue, of course, is whether the factors, which have kept IT energy demand relatively constant, have run their course and demand will now begin to increase following the forecast growth in demand for information services. Novel, computationally intensive services like Bitcoin are wild cards. At a minimum this is an area where energy efficiency innovation will play a critical role.

4.3. *Electrification*

The recent Intergovernmental Panel on Climate Change (IPCC) report concludes that keeping greenhouse warming to tolerable levels will require reducing net greenhouse gas emissions to (or even below) zero per cent by 2050.[ai] This goal will be difficult to meet if fossil fuels are used where the emissions cannot be captured and permanently sequestered. Carbon

---

[ah] Google Data Centers, "Efficiency: How we do it". https://www.google.com/about/datacenters/efficiency/internal/.

[ai] IPCC, "Global warming of 1.5°C", Intergovernmental Panel on Climate Change, 2018. https://report.ipcc.ch/sr15/pdf/sr15_spm_final.pdf.

capture is practical only in large industrial facilities and electric generators. This means that fuel use by most buildings, vehicles and industrial facilities must be replaced by systems using electricity, biofuels, or possibly synthetic fuels made from electricity. Renewable fuels include hydrogen made from low-emission electric generation and synthetic fuels made from the renewably produced hydrogen and carbon dioxide captured from the air. Recent analysis suggests that in most cases they are unlikely to be economically competitive because applications use electricity directly (i.e., electric vehicles, heat pumps).[17] Fossil fuels could indirectly be used for these distributed systems since they can be used to generate electricity where carbon capture and sequestration is feasible. It is also theoretically possible to offset emissions from fossil fuels with equipment capable of removing carbon dioxide from the atmosphere, a strategy that the IPCC recognizes is filled with uncertainty. The technical challenges are enormous and aggressive policies would need to be implemented to make investments in such systems profitable to investors.

Economy-wide shifts from direct use of fossil fuels to electricity would redefine priorities, focusing work on electric technologies. Policies driving such a shift could benefit by replacing older, less efficient systems with new equipment that meet high efficiency standards. This shift would have a major impact on electric utilities by adding to electricity demand, shifting peak loads, and by offering unprecedented ability to manage electric demand in ways that could minimize overall efficiency and costs of producing and using electric energy.[18]

Inexpensive ways to store electricity would greatly expand options and many new technologies (e.g., advanced building controls with thermal and electric vehicle batteries) can provide storage options.

How feasible is it to replace the distributed use of fossil fuels? Taken together, fuels used for purposes other than electricity generation account for nearly 60% of fuel use worldwide (Table 2).

- There is already a significant market for *buildings* powered only by electricity, since nearly all the fuel used in buildings involves space heating and hot water. The efficiency of heat pumps, particularly those designed for colder climates, will continue to increase. Sales of electric heat pumps in 2017 reached 3.7 GW.[aj] It is also possible to provide electric power and heat to buildings using hydrogen powered fuel

---

[aj] IEA, 2018b.

Table 2.   On-site fossil fuel consumption as a percentage of total energy consumption in 2015.

	World (%)	OECD (%)
Residential and Commercial	8.1	10.0
Transportation fuel	19.0	24.6
Industrial fuel	32.2	23.9
Total	59.2	58.6

Source: IEA Data online, 2018.

cells.[ak] The impact of a large-scale shift from fossil fuel heating systems in buildings to electricity on electric generation and distribution companies will depend critically on climate. In California, for example, cooling loads are well correlated with renewable resources while heating loads are not. While climate change would increase cooling demand and decrease heating demand, the net effect on electricity consumption would be small. The shift to electric heating, however, could be significant. Less than 15% of residential buildings in California now use electric heat and a shift to total electric heating would increase demand for electricity by more than 13%.[19]

- It is also feasible to replace most *highway vehicles* with electric vehicles (EVs) or hydrogen powered fuel cells. Eight countries have announced impending bans on the sales of conventional gasoline and diesel cars by 2040, and 19 cities have established restrictions on internal combustion engines (ICE), which will come into full effect in the same year. However, some will come into effect by 2025. Electric vehicle sales are only about 1% of vehicles on the road, though sales are growing quickly and there are now more than 3 million electric cars on the road and 370,000 electric buses. Three hundred million electric 2 and 3-wheel vehicles are now operating in China.[al] Elsewhere, EV sales reached 40% of sales in Norway.[am] Electric options are certainly possible for some types of *aircraft, heavy trucks and off-road vehicles* such as construction equipment, but it seems likely that these vehicles will need to pay a

---

[ak]US DOE, "State of the states: fuel cells in America 2017", US Department of Energy, 2018. https://www.energy.gov/sites/prod/files/2018/06/f52/fcto_state_of_states_2017_0.pdf.

[al]https://webstore.iea.org/download/direct/3007.

[am]IEA, "Tracking Clean Energy Progress", International Energy Agency.

Table 3:   Industrial energy use in Europe (8.9 EJ total).

Temperature of process (°C)	Energy used (EJ) (%)
>500	42.9
100–500	29.3
<100	9.4
Space heat	14.3
Process cool	3.6
Space cool	0.5
Total	100.0
Low temperature	27.0
Heating/cooling	14.8

premium to have access to liquid fuels from biomass or other non-fossil sources.

- *Industrial* fuel use presents unique challenges because of the diversity of processes involved. About a quarter of industrial energy use in Europe is used for heating and cooling at temperatures only reachable by commercial heat pumps (Table 3).[20] Steel, cement, petrochemicals and other processes require temperatures above 100°C — temperatures that cannot be supplied with contemporary heat pumps. However, electric processes can be used for at least some of these applications, for example, electric arc furnaces used in steel production. Hydrogen produced by electrolyzing water can be used for the direct reduction of iron ore.[21] Electricity can be used to produce ammonia from atmospheric nitrogen at ambient temperatures and pressures.[22] Electric powered additive manufacturing can replace traditional forging methods.

The reduction in the mass of materials used per dollar of GDP in modern economies ("demassification") may lead to declining demand for materials that require high temperatures.[23] Energy efficiency in vehicles and other systems often results from strategies to reduce weight. These trends are likely to continue as developing economies mature. Petrochemical energy use will decline with the decline in use of oil and coal needed to meet climate goals.

Countering these trends, however, is the real possibility that if action on climate mitigation is delayed, adapting to climate change may require massive investments in new infrastructure as sea-level changes and unbearable heat make some highly populated areas uninhabitable.

5. Looking Forward

Two major recent studies[an],[ao] of ways to meet the goals of the Paris agreement on climate change call for changes in global energy infrastructure that are "unprecedented in scale and speed". Meeting the goal of keeping global temperature increases below 1.5 or 2°C requires investments in new sources of clean energy, energy efficiency and technologies for removing carbon dioxide from the atmosphere. The discussion in this volume explores the large gains that can be made by investments in innovative energy efficiency technologies and the new business models built around them. It often suggests that the opportunities have been underestimated. The modularity and affordability of efficiency investments and the fact that they do not involve complex siting or land use challenges make it far more likely that efficiency investments can meet this challenge, given adequate policy support.

The challenge of massive investments in new clean-energy production has been widely discussed but new studies argue that meeting climate goals will also require large investments in biomass and technologies of "carbon dioxide removal". The reports recognize that these strategies are fraught with risks — environmental, technical (most carbon dioxide removal technologies are unproven) and management — and uncertainties. Carbon dioxide removal investments make no sense without a global incentive such as a price on carbon and many would require international acceptance and management. Large changes in land management would trigger political challenges and would almost certainly also require complex international agreements.

The need for such drastic measures can be greatly reduced if energy efficiency strategies can exceed conventional expectations. One of the IPCC scenarios, drawing on the work of Grubler and colleagues,[24] provides a concrete demonstration of what is possible. Low global energy demand also alleviates some of the pressure to make massive investments in new energy supply technologies needed to meet environmental goals in solar, wind, biomass and others forms of energy demands.

The most compelling reason to expect energy efficiency to play an immediate and major role in meeting global environmental goals is that the innovations driving the efficiency gains are almost always identical,

---

[an]IPCC, 2018.
[ao]IEA, "Energy technology perspectives", International Energy Agency, 2017.

with investments needed to drive continued improvements in global living standards. Unlike investments like carbon removal, efficiency investments lead to more comfortable living spaces, faster and safer ways to get to a desired destination, and smarter and cheaper manufactured products. They drive investments in system-wide efficiencies made possible by increasingly powerful networks of sensors, communication and analytics, many of them made possible by electrification. Economic growth is increasingly driven by the collection and manipulation of information and associated electronic equipment.

The *Time to Choose* report introduced the meme that "changing massive energy systems was like turning a supertanker" — turning at a glacial pace. The long lifetimes of vehicles, buildings and industrial equipment will clearly limit the pace of change, but the link between energy efficiency innovation and the momentum of modern economies provides reason to believe that the rate of change can be much faster than it has been in the past. The modularity of the investments provide much more agility than projects that may take decades to plan and execute. The rate of change in mobility driven by Uber and similar firms was completely unexpected. New vehicles will probably not enter the market as fast as cell phones but with increasing electrification and digitization. It is certainly possible that the pace will quicken. Automated, connected devices allow global upgrades to be installed quickly and efficiently.

While market forces are driving investments in energy efficiency, wise public intervention will be essential to achieve the speed and scale required for the major transformations needed to achieve economic and environmental goals. The most obvious defect in contemporary energy markets is that energy prices do not include the cost of environmental damage and other forms of damage resulting from their use.[ap] Taxing carbon is the most straightforward approach to ensuring that the full costs of energy use are recognized by markets, but the political difficulty of imposing them is apparent. Regulations, labels, information programs, subsidies and many other policy interventions can be effective proxies for price signals, and some of these measures may be needed even if appropriate taxes are in place. Taxes are, however, the best way to minimize the "take-back" effect created when efficiency investments lower the cost of a service like driving.

[ap] National Research Council, *The Hidden Costs of Energy* (The National Academies Press, Washington, 2010). https://www.nap.edu/catalog/12794/hidden-costs-of-energy-unpriced-consequences-of-energy-production-and.

This thereby encourages an increase in driving, which negates the environmental benefits of efficiency.

Public regulation will also be needed to facilitate the emergence of many of the system-wide efficiency transformations discussed earlier. New mobility systems will require standardized methods for vehicles to communicate with each other and with the highway system. Next-generation public-private partnerships in mobility will require agreed ways to manage scheduling and payment data. System costs can only be minimized if control over equipment that uses electricity is intelligently integrated with systems for generating, storing and delivering electricity. Minimizing costs will require agreement on how price and other signals are communicated. Urban design and operations deeply shape opportunities for efficient transportation and building energy use. While commercial actors will lead to the development of the standards needed for these new systems to operate, most will also require some form of public intervention, if only to ensure that public safety, services, and other benefits and costs are considered. Caution is always needed since both premature regulatory intervention and sluggish, inadequate intervention can both frustrate innovation.

Energy prices set below their real cost to society, coupled with significant uncertainty about the future of public programs, has led to a consistent underinvestment in energy efficiency research. In spite of decades of innovation, buildings, mobility and product fabrication still operate far from the theoretical efficiency limits. Public investment in basic and applied energy research remains essential.[25] Since many of the innovations that contribute to energy efficiency also contribute directly to other public goals such as improved transportation, housing, productivity, employment, environmental and basic scientific knowledge, efficiency research must be much more carefully integrated into broader national research agendas.

There is no discounting the magnitude of the challenge we face: finding a way to build a world where nine billion people can live comfortable, fulfilling lives without compromising the future of the global environment. It is clear that creative strategies for advancing energy efficiency are central to the solution.

References

1. I.S. Wing and R. Wickhaus, "The decline in US energy intensity: its origins and implications for long-run CO_2 emission projections", MIT Economics Department and Joint Program on the Science & Policy of Global Change. https://economics.mit.edu/files/2624.

2. W. Rosen, *The Most Powerful Idea in the World* (University of Chicago Press, Chicago, 2010).

3. D. Gilbert, "On the progressive improvements made in the efficiency of steam engines in Cornwall, with investigations of the methods best adapted for imparting great angular velocities", *Philosophical Transactions of the Royal Society of London* **120**(1830) 121–132. https://royalsocietypublishing. org/doi/10.1098/rstl.1830.0010. Accessed: 10 August 2020.

4. Energy Policy Project of the Ford Foundation, *A Time to Choose, A Time to Choose: America's Energy Future* (Ballinger Publishing Company, Cambridge, 1974).

5. R. Yin, D. Black, M.A. Piette and K. Schiess, "Control of thermal energy storage in commercial buildings for california utility tariffs and demand response", LBNL-1003740, Lawrence Berkely National Laboratory, 2015. https://eta.lbl.gov/sites/all/files/publications/control_of_thermal_energy_ storage_in_commercial_buildings_for_california_utility_tariffs_and_ demand_response_lbnl-1003740.pdf.

6. R. Sheehan, "Deployment of the ITS and Operations in the US", ITS World Conference, Bordeau, France, 2015. https://www.its.dot.gov/presentations/ world_congress2015/wc2015_PR12_%20ITSOperations.pdf.

7. J.S. Alonso-Mora, "On-demand high-capacity ride-sharing via dynamic trip-vehicle assignment", *Proceedings of the National Academy of Sciences of the United States of America*, 2017, pp. 462–467. https://www.pnas.org/ content/114/3/462.

8. L. Fulton, "Three revolutions in urban transportation: how will these affect the costs of trips", STEPS Symposium, UC Davis, 12 July 2017. https:// steps.ucdavis.edu/wp-content/uploads/2017/12/FULTON-3R-scenarios-for-STEPS-7dec17-final.pdf.

9. Z. Wadud, D. MacKenzie and P. Leiby, "Help or hindrance? The travel and carbon impacts of highly automated vehicles", *Transportation Research Part A: Policy and Practice* **86** (2016) 1–18. https://doi.org/10.1016/j.tra.2015.12.001.

10. J.H. Gawron, G.A. Keoleian, R.D. De Kleine, T.J. Wallington and H.C. Kim, "Life cycle assessment of connected and automated vehicles: sensing and computing subsystem and vehicle level effects", *Environmental Science and Technology* **52** (2008) 3249–3256.

11. M. Levinson, *The Box* (Princeton University Press, 2016).

12. Pitney Bowles, "Parcel shipping index". https://www.pitneybowes.com/ content/dam/pitneybowes/us/en/shipping-index/pitney-bowes-parcel-ship-ping-index-infographic-2018.jpg.

13. T. Hettesheimer, S. Hirzel and H. B. Roß, "Energy savings through additive manufacturing: an analysis of selective laser sintering for automotive and aircraft components", *Energy Efficiency* **11** (2018) 1227–1245.

14. J. Dedrick, G. Linden and K.L. Kraemer, "We estimate China only makes $8.46 from an iPhone — and that's why Trump's trade war is futile", *The*

Conversation, 6 July 2018. https://theconversation.com/we-estimate-china-only-makes-8-46-from-an-iphone-and-thats-why-trumps-trade-war-is-futile-99258.

15. J. Malmodin and D. Lundén, "The energy and carbon footprint of the global ICT and E&M sectors 2010–2015", *Sustainability* **10** (2018) 3027. https://www.mdpi.com/2071-1050/10/9/3027/htm.

16. J.G. Koomey, S. Beard, M. Sanchez and H. Wong, "Implications of historical trends in the electrical efficiency of computing", *IEEE Annals of the History of Computing*, **33** (2011).

17. Aas, Dan, Amber Mahone, Zack Subin, Michael Mac Kinnon, Blake Lane, and Snuller Price. 2020. The Challenge of Retail Gas in California's Low-Carbon Future: Technology Options, Customer Costs and Public Health Benefits of Reducing Natural Gas Use. California Energy Commission. Publication Number: CEC-500-2019-055-F. https://ww2.energy.ca.gov/2019 publications/CEC-500-2019-055/CEC-500-2019-055-F.pdf.

18. A. Ovalle, A. Haly and B. Seddik, *Grid Optimal Integration of Electric Vehicles: Examples with Matlab Implementation* (Springer, 2018). https://doi.org/10.1007/978-3-319-73177-3.

19. B. Tarroja *et al.*, "Translating climate change and heating system electrification impacts on building energy use to future greenhouse gas emissions and electric grid capacity requirements in California", *Applied Energy* **225** (2018) 522–534.

20. M. Rehfeldt, T. Fleiter and F. Toro, "A bottom-up estimation of the heating and cooling demand in European industry", *Energy Efficiency* **11** (2018) 1057–1082.

21. J. Wiencke, "Electrolysis of iron in a molten oxide electrolyte". https://link.springer.com/article/10.1007/s10800-017-1143-5.

22. R. Lan, J.T.S. Irvine and S. Tao, "Synthesis of ammonia directly from air and water at ambient temperature and pressure", *Nature Scientific Reports*, 2013. http://doi.org/10.1038/srep01145.

23. J.H. Ausubel and P.E. Waggoner, "Dematerialization: variety, caution, and persistence", *Proceedings of the National Academy of Sciences of the United States of America* **105** (2008) 12774–12779. https://doi.org/10.1073/pnas.0806099105.

24. A. Grubler *et al.*, "A low energy demand scenario for meeting the 1.5°C target and sustainable development goals without negative emission technologies", *Nature Energy* **3** (2018) 515–527.

25. K.S. Gallagher, A. Grubler, L. Kuhl, G. Nemet and C. Wilson, "The energy technology innovation system", *Annual Review of Environment and Resources* **37** (2012) 137–162.

© 2021 World Scientific Publishing Company
https://doi.org/10.1142/9789811217869_0002

Chapter 2

Global Opportunities and Challenges in Energy and Environmental Issues in the Buildings Sector

Mary Ann Piette[*,§], Rick Diamond[*,¶], Stephen Selkowitz[*,‖],
Stephane de la Rue du Can[*,**], Tianzhen Hong[*,††], Kaiyu Sun[*,‡‡],
Paul Mathew[*,§§], Iain Walker[*,¶¶], Alan Meier[*,‖‖], Erik Page[†,***],
Jessica Granderson[*,†††], Nan Zhou[*,‡‡‡] and Peter Alstone[‡,§§§]
[*]Lawrence Berkeley National Laboratory, Berkeley, CA 94720, USA
[†]Erik Page and Associates, California, USA
[‡]Humboldt State University, Arcata, CA 95521, USA
[§]mapiette@lbl.gov
[¶]rcdiamond@lbl.gov
[‖]seselkowitz@lbl.gov
[**]sadelarueducan@lbl.gov
[††]thong@lbl.gov
[‡‡]ksun@lbl.gov
[§§]pamathew@lbl.gov
[¶¶]iswalker@lbl.gov
[‖‖]akmeier@lbl.gov
[***]erik@erikpage.com
[†††]jgranderson@lbl.gov
[‡‡‡]Nzhou@lbl.gov
[§§§]peter.alstone@humboldt.edu

1. Introduction and Global Energy Use in the Buildings Sector

As the United States (US) and other countries evaluate energy consumption trends, the opportunities to improve energy efficiency and reduce greenhouse gas emissions from the building sector are critical elements of national energy policies and strategies. Although the building sector hosts an incredibly diverse set of energy end uses, an exciting array of new technologies, business opportunities and process improvements are continuing to show great promise. Energy efficiency programs sponsored by electric utilities provide robust demand side management investments, where reducing a kilowatt-hour of electricity costs only 2.5 cents.[1]

This chapter provides an overview of the global opportunities and challenges in energy and environmental issues in the buildings sector. It begins with an overview of the buildings sector in the US and the total energy use and costs to operate these buildings, then introduces the current and forecasted energy use in both developed and developing countries, with an emphasis on China and India. Half of all the buildings being constructed today are in China, thus the opportunity to influence energy use to support low-energy buildings there is critical. India also continues to see high building construction rates, and the growing use of air conditioning in India and other developing countries is one of the critical challenges of our time. This chapter includes some examples of achievements in building energy efficiency from the last decade, providing examples of technologies and policies that have brought about huge improvements in both whole building and appliance energy use.

The greater levels of renewable energy systems being integrated into our energy supply are challenging utility energy efficiency programs. Hourly load shapes are more important than ever. While energy efficiency has dominated utility demand-side portfolios, new distributed energy resources are emerging with the need for new valuation and investment structures. This chapter describes how the next generation of efficient buildings must have controls that can integrate with the electric grid, either a microgrid or the larger macrogrid.

This chapter is organized as follows. Section 1 provides an overview of current global energy use in the buildings sector, and how it is changing. Section 2 outlines how energy is used in buildings in the US, providing an overview of the type of energy used for residential and commercial buildings, along with a description of how much energy is used by different end uses. Section 3 describes numerous examples of the diverse and robust opportunities to reduce energy use in buildings, with a summary of technologies in facades; heating, ventilation and air conditioning (HVAC); lighting; and other technologies. Section 4 describes energy use in China and India, which is critically important for global energy projections. Section 5 describes how the electric grid is changing with the introduction of greater levels of variable renewable energy. This change introduces new challenges to ensure that flexible building loads and distributed energy resources are integrated in a way that supports a low-carbon energy economy.

1.1. *Overview of the global buildings sector*

The buildings sector is the largest source of primary energy consumption (40%) and ranks second after the industrial sector as a global source of direct and indirect carbon dioxide (CO_2) emissions from fuel combustion, emitting 10.1 gigatons of carbon dioxide ($GtCO_2$) in 2016 (Fig. 1). Building carbon dioxide emissions have increased by 10% in 10 years (2006–2016) and largely are related to electricity consumption. Indirect carbon dioxide emissions from electricity consumption represent 73% of total building emissions.

The International Energy Agency (IEA) reports that energy demand from buildings and building construction continues to rise, driven by improved access to energy in developing countries, greater ownership and use of energy-consuming devices, and rapid growth in global buildings floor area, at nearly 3% per year.

Globally, electricity use has increased in both commercial and residential buildings over the past 20 years (Fig. 2). This growth is particularly notable in commercial buildings, where the share of electricity consumption grew from 17% of final energy consumed in 1971 to 51% in 2016.[2] This growth has been driven by an increasing penetration of electrical equipment for space cooling, air ventilation and lighting, as well as growth in office equipment. Residential sector electricity consumption accounts for

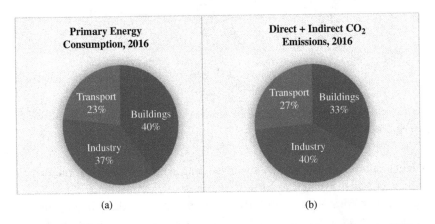

(a)   (b)

Fig. 1. (a) Primary energy consumption and (b) carbon dioxide emissions in 2016, by sector. Energy and emissions from the power sector were reallocated to the end use sectors according to their electricity consumption.

Source: IEA.[2]

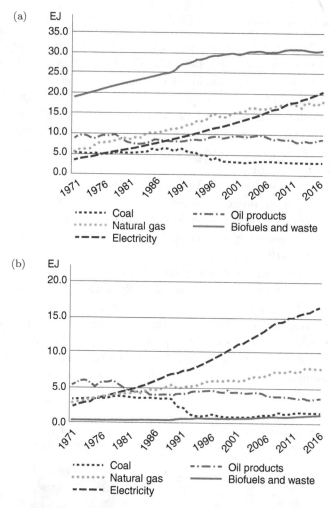

Fig. 2.   Global energy use over time for residential (a) and commercial buildings (b).
Source: IEA.[2]

a much smaller share (24%). In developing countries, a large quantity of
energy is still used in the form of biomass, which accounted for 35% of
final energy consumption in 2016, according to the IEA. If only modern
forms of energy are accounted for, electricity represents almost half of the
final energy consumption of residential buildings. To date, electricity use
in the buildings sector has largely been attributable to developed countries. However, electricity use is growing more rapidly in developing countries as rural populations move to cities and economies become more

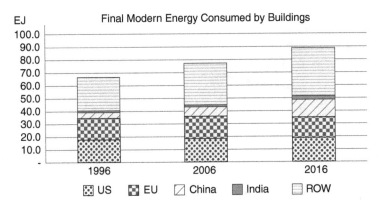

Fig. 3. The world's total energy use from buildings showing the US, European Union, China, India, and the rest of the world (ROW).

Source: IEA.[2]

service-based. According to a recent United Nations study,[3] the world's population living in urban areas is expected to increase from 55% today to 68% by 2050. With the overall growth of the world's population, this translates into 2.5 billion new urban citizens by 2050, 90% of whom will live in Asian and African cities.

Regionally, the US is the largest source of modern energy consumption in buildings (representing 21%), followed by the European Union (EU) (18%) and then China (15%). These three regions, along with India (3%), represent 58% of the total modern energy consumption (Fig. 3).

2. The US Building Sector

Buildings in the US account for more than 40% of its total energy use — along with the associated greenhouse gas (GHG) emissions — and more than 7% of all the US's electricity use. This energy is used to provide the heating and cooling for human comfort in residential and commercial buildings, as well as all the services needed in these and in industrial buildings. To understand the demand for energy services in the US's building sector, this chapter examines specific characteristics of the US commercial and residential building stock viz. the size, age, type and other factors that impact the energy used in buildings. It then reviews the total energy use, energy use intensity and the end-use consumption in both the commercial and residential sectors.

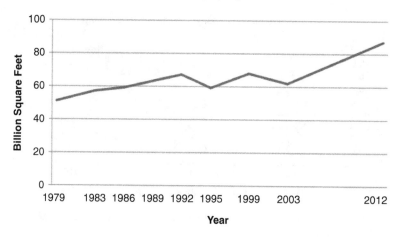

Fig. 4. Growth in the US commercial building floor area (1979–2012). *Source*: EIA (2016, Table B-1).

2.1. *US commercial sector building characteristics*[a]

The best source for information on the US commercial building stock is the US Department of Energy's *Commercial Buildings Energy Consumption Survey* (CBECS), which defines the US commercial sector as "buildings greater than 1,000 square feet that devote more than half of their floor-space to activity that is not residential, manufacturing, industrial, or agricultural".[4]

The most recent CBECS (2012 data) estimates that there were 5.6 million commercial buildings in the US, comprising 87 billion square feet of floorspace. This level represents a 14% increase in the number of buildings and a 21% increase in floorspace since 2003 (Fig. 4). A growing population has led to a need for more buildings, and the changing needs and wants of consumers has led to larger buildings.

2.1.1. *Size of US commercial buildings*

Although there are relatively few large buildings in the US, defined as more than 100,000 square feet (10,000 square meters [m^2]) of floorspace,

---

[a]Summary from EIA, 2016, the 2012 Commercial Buildings Energy Consumption Survey (CBECS), 4 March 2015. https://www.eia.gov/consumption/commercial/reports/2012/buildstock/index.php.

they account for more than one-third of total commercial building floor-space.[4] (While commercial buildings are often pictured as a skyline of towering buildings, in reality, the vast majority of US commercial buildings are relatively small. About half of the commercial buildings are smaller than 5,000 square feet (500 m^2) in floor area, and nearly three-fourths are smaller than 10,000 square feet (1,000 m^2) in floor area.

2.1.2. *Year constructed for US commercial buildings*

Commercial buildings remain in use for many decades. Roughly 12% of commercial buildings (comprising 14% of commercial floorspace) were built since 2003, and nearly half of all US commercial buildings were constructed before 1980; the median age of buildings in 2012 was 32 years.

2.1.3. *Energy consumption in US commercial buildings*

In 2012, US commercial buildings used 6.9 quadrillion Btu (quads) of total site energy: 4.2 quads of electricity, 2.2 quads of natural gas and 0.1 quads of fuel oil. Electricity and natural gas usage increased by 19% and 7%, respectively, since 2003, while fuel oil and district heat usage decreased by 41% and 46%, respectively. Overall, total energy usage in commercial buildings increased 7% since 2003 (Fig. 5).

The slower growth in commercial building energy demand since 2003 may be explained in part, by newer construction that is built to higher energy performance standards, occupied by less energy-intensive building activities, and more often built in temperate regions. The improved efficiency of key energy-consuming equipment is also decreasing demand. Since 2003, for example, space heating and lighting are each down by 11% in their share of energy use in buildings.

According to CBECS, the increase in the amount of electricity consumed in US commercial buildings is consistent with the adoption of new types of electronic equipment, as well as the increased use of existing technologies such as computers and servers, office equipment, telecommunications equipment, and medical diagnostic and monitoring equipment. In addition to electricity consumed directly by the equipment, many of these electronics require additional cooling, humidity control, and/or ventilation equipment, which also increases electricity consumption.

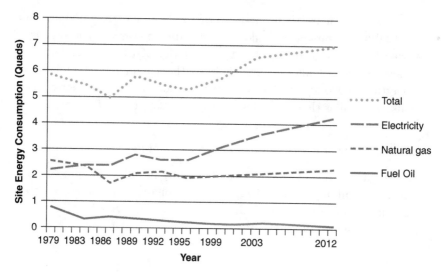

Fig. 5. US commercial building site energy consumption (quads) by fuel type, 1979–2012.

Source: EIA, 2016.[4]

2.1.4. *Energy intensity in US commercial buildings*

The average total (site) energy used per square foot (ft^2) of commercial buildings has decreased since 2003, from 91 kBtu per ft^2 to 80 kBtu per ft^2 (Fig. 6). The average electricity use per square foot remained about the same since 2003, but decreased for natural gas. The decrease in natural gas energy intensity is likely related to federal equipment standards over that time period, and the warmer-than-average winter months of the CBECS survey year in 2012.

Commercial building energy use can be divided into separate end uses. The largest end-use share of total energy (site) in US commercial buildings is for space heating, which accounted for 25%, followed by the "other" category, which represented 12%. Lighting, cooling, ventilation and refrigeration each accounted for 10% of total energy use (Fig. 7).

2.1.5. *Electricity consumption in US commercial buildings*

Total electricity consumption in US commercial buildings has almost doubled between 1979 and 2012, increasing from slightly more than

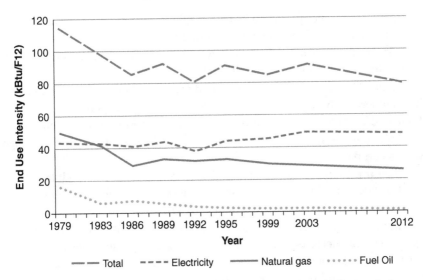

Fig. 6. US commercial building energy use intensity (kBtu/ft^2) by fuel type, 1979–2012.

Source: EIA (2016, Table E-1).

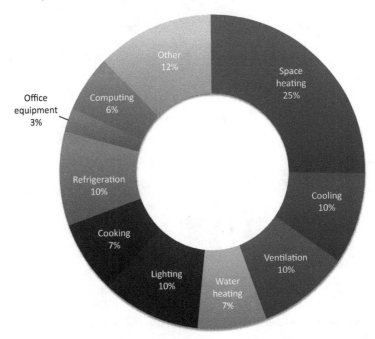

Fig. 7. Total energy by end use in US commercial buildings, 6,963 trillion Btu (2012).

Source: EIA (2016, Table E-5).

2.2 quads (site energy) in 1979 to 4.2 quads (site energy) in 2012. The end-use breakdown of commercial building electricity use shows the largest end use is for lighting (17%), followed by ventilation and refrigeration (each 16%), and cooling (15%) (Fig. 8).

Since 2003, the share of energy used for space heating and for lighting in commercial buildings have each decreased by 11 percentage points. Increased shares are estimated for cooking, refrigeration, computing (including servers) and "other", each of which was up by 4 percentage points compared to 2003.

The total amount of energy used for lighting has decreased 46% from 2003 to 2012 (Fig. 9), a change in large part due to the increasing use of fluorescent and light-emitting diode (LED) lamps as replacements for incandescent bulbs. As noted above, the total amount of energy used for space heating decreased 26% from 2003 to 2012, likely because of the warmer-than-average winter during the CBECS reference year (2012) and to improve federal equipment standards.

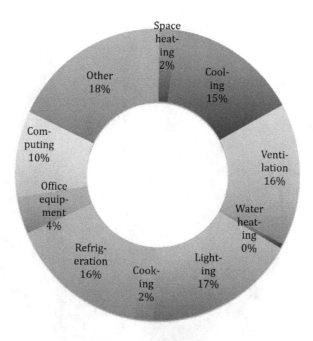

Fig. 8. US commercial building electricity end-use consumption, 2012. 1,243 billion kWh (2012).

Source: EIA (2016, Table E-5).

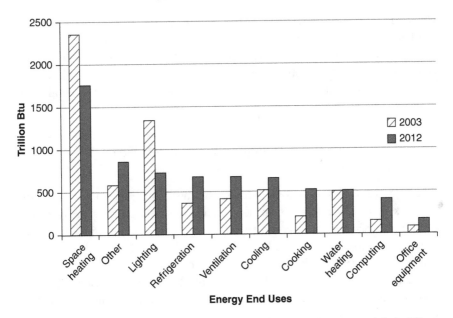

Fig. 9. Change in end-use energy consumption for US commercial buildings (2003–2012).
Source: EIA (2016, Table E-1).

2.2. US residential sector building characteristics[b]

Residential buildings account for 55% of the energy used in buildings in the US. The 2015 Residential Energy Consumption Survey (RECS) describes the characteristics that contribute to the energy consumption in US homes, and highlights the variability in the way energy is used across the nation's 118 million households. Household energy use varies by geographic location, structural features, equipment choices and energy sources used. Electric heat pumps, for example, are well suited for heating in areas where winters are relatively mild. Of the 11.8 million households in 2015 that used electric heat pumps, 8.4 million (71%) were in the South. The high cost of electric heating in colder climates has often limited the use of heat pumps and other electric equipment in

---

[b]Summary from EIA, 2018 Residential Energy Consumption Survey (RECS). https://www.eia.gov/consumption/residential/data/2015/.

those areas. People moving to warmer climates has also increased household use of air conditioning. In 2015, more than 76 million households (64%) used a central air-conditioning system, an increase from 66 million households (59%) in 2005. Space conditioning (heating and air conditioning) accounts for about half of all energy consumed in the residential sector.

2.2.1. *Impact of age and location of household on energy consumption*

The Hot-Humid climate zone, which stretches from Florida to southeastern Texas, has seen a surge in new home construction over the past few decades, particularly compared to the rate of construction in colder areas. While all homes in the Hot-Humid zone make up 19% of all US households, nearly 28% of the homes built since 2000 are in that area. Given the warmer weather and predominance of the use of electricity in home cooling, average household electricity consumption is higher in that climate zone than in other areas of the country.

New homes in the US are characterized by different structural features than older homes. According to the 2015 RECS, more than half of all new homes have high ceilings, compared with 30% of homes built before 2000.

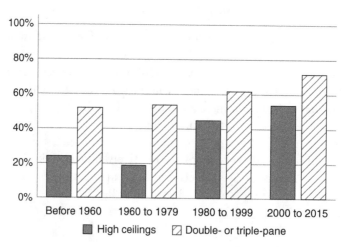

Fig. 10. Newer houses have higher ceilings and more efficient windows.
Source: EIA (2018a, Table HC-3).

High ceilings tend to increase heating loads. At the same time, double- and triple-pane windows, which reduce the loss of heated and cooled air, are also more prevalent in new homes (Fig. 10).

2.2.2. *Household fuel type*

Many US households rely on a variety of sources to meet their home energy needs. In 2015, 75% of US households used more than one source of energy, 66% used electricity and one other fuel, and 9% used three or more fuel sources. A quarter of the total number of households used only electricity. The combination of energy sources used in households takes many forms, from widespread use of natural gas and petroleum fuels in cold regions to less frequent use of wood in a fireplace as a secondary heat source. When classifying homes by type of construction, single-family detached homes were the least likely to be all electric (18% in 2015), while mobile homes were the most likely (44% in 2015).

Trends in space heating are partly responsible for the increase in all-electric homes in the US. In 2015, electricity was used as the main source of energy for space heating in 36% of homes that were heated, an increase from 2009 that continues a long-term trend. Natural gas continued to be the most common energy source for main space heating, with 51% of heated homes using natural gas in 2015. The 118 million US households used 77 million Btu on average in 2015, with the average annual household expenditure for energy being just under $2,000 per year. Of this total (site) energy consumption, 47% was provided by electricity, 44% by natural gas, and 9% from fuel oil and propane. Heating, ventilating and cooling account for more than half of the total household energy use (54%) and the rest is split into several dozen end uses (Fig. 11).

Figure 12 shows the breakdown in residential electricity use. The largest end use of electricity in US households is for space cooling (22%), followed by space heating (17%) and then water heating (14%).

The average US household spent $1,856 on home energy bills in 2015. Although air conditioning accounted for 12% of the total household energy costs (and 17% of electricity expenditures) at the national level, some regions use much more air conditioning. In the hot-humid region, where air conditioning was used by 94% of households, air conditioning made up 27% of home energy expenditures. By comparison, in the marine region, where nearly half of the households did not use any air conditioning, air conditioning made up just 2% of home energy expenditures.

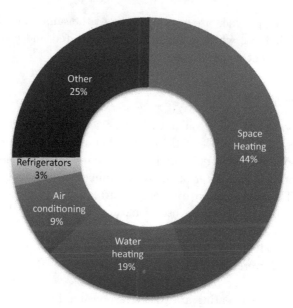

Fig. 11.   Residential energy end uses (site energy), 2015.
Source: EIA (2018a, Table CE-3.1).

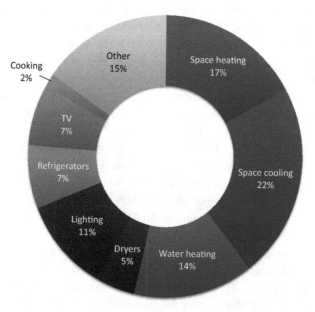

Fig. 12.   Residential electricity consumption by end use, 2015.
Source: EIA[5] (2018a, Table CE-5.1a).

3. Building Technologies and Opportunities to Reduce Energy Use

This section provides an overview and examples of emerging technologies that can greatly reduce energy use in buildings. It begins with a discussion of energy efficiency in windows and façades, followed by a review of new HVAC technology, and interior lighting technologies. This section also includes a discussion of energy use of miscellaneous equipment, which is a growing load in many buildings. We present an overview of simulation tools for evaluating energy use in both new and existing buildings. This is followed by sections on energy use issues in residential and commercial buildings, including energy use rating systems, indoor air quality issues and benchmarking.

3.1. *Windows, skylights, glazings and building facades*

Heating and cooling are the dominant uses of energy in buildings (Sections 2.1 and 2.2). Windows frequently have heat loss and heat gain rates that are 5 to 50 times higher than opaque insulated walls or roofs, as frequently observed qualitatively in long-wave IR imagery (Fig. 13). The initial design response — embodied widely in building standards since the 1970s — has been to restrict window area while mandating modest thermal improvements, such as moving from single to double glazing.

Fig. 13.   Infrared image of a building showing thermally poor windows.

As with many other building elements, there are tremendous technical and market opportunities to improve the energy performance of windows,[c] and great progress has been made over the past 50 years as outlined below. Less well known or understood, however, is the technical potential to turn almost any window into a net positive energy supplier to the building in virtually any climate. This requires shifting one's perspective from minimizing losses in the traditional static role of the window to reconceptualizing the window as a dynamic, adaptive, intelligent building system with the ability to (1) minimize thermal losses and maximize useful solar gain in winter, (2) minimize gains and solar-driven cooling loads in summer while admitting useful light, (3) admitting glare-free daylight year round, (4) varying the permeability of the window system to optimize free-cooling with natural ventilation on a seasonal basis, (5) adding thermal storage to manage short-term time-dependent flows, and (6) converting the window to a power-generating element by adding transparent or semi-transparent power generating capabilities to the window system.

Successful achievement of these technical potentials will not only help support the building level drive to zero net or positive energy balance[6] but also provides a degree of design freedom to architects and engineers that will transform homes and workplaces into more inviting, healthier, and more productive living and working environments.

The potential changes in building sector energy use by employing these new technologies are significant. Table 1 examines the impacts of current window energy use in both the residential and commercial sectors with the "installed" base windows. In comparison, Tables 2 and 3 show the calculated potential savings if multiple scenarios with improved window products were implemented.

In both sectors the technical potential of 2.25 quads and 2.62 quads exceed the estimated end use impacts of 2.24 quads and 1.39 quads of existing stock. While this includes the daylighting savings in the commercial sector, this assessment does not include any credit for power generation in the window with the use of photovoltaic devices.

Technical potential studies quantify theoretical savings but rarely define what is achievable in real markets due to many issues that have been studied by the energy efficiency community over the past 50 years.

---

[c]We use the term "windows" here to refer to the full range of glazed building openings in all building types, e.g., windows and patio doors in residential buildings, punched windows, curtain walls and double envelope facades in commercial buildings, and skylights and roof glazings in all building types.

Table 1. Total window energy impacts.

	Total annual HVAC consumption	Window percent of HVAC consumption, no infiltration	
	Quadrillion primary Btu (quads)	% of total	Window (quads)
Residential heat	6.90	19	1.30
Residential cool	2.41	39	0.94
Residential total	9.31	24	2.24
Commercial heat	2.45	35	0.85
Commercial cool	1.90	28	0.54
Commercial total	4.35	32	1.39

Source: Arasteh *et al.*[148]

Table 2. Annual energy savings potential of residential window technologies.

	Energy savings over current stock (quads)		
Window type	Heating	Cooling	Total
Sales (business as usual)	0.49	0.37	0.86
ENERGY STAR (low-E)	0.69	0.43	1.12
Dynamic low-E	0.74	0.75	1.49
Triple-pane low-E	1.20	0.44	1.64
Mixed-triple, dynamic	1.22	0.55	1.77
High-R superwindow	1.41	0.44	1.85
High-R dynamic	1.50	0.75	2.25

Notes: "Dynamic": Glazings with switchable optical properties. "High-R Superwindow": R 8-10 window, achieved with triple glazing, vacuum glazing, aerogel, etc.

In the case of windows, the biggest challenges include the increased cost of high performance technologies, the long life and slow turnover of windows in buildings, supply chain complexities, high cost investments for new glass and coating manufacturing capability, design optimization for climate and orientation, installation and maintenance, integration with

Table 3. Annual energy savings potential of commercial window technologies.

Window type	Energy savings over current stock (quads)			
	Heating	Cooling	Lighting	Total
Sales (business as usual)	0.03	0.17	—	0.20
Low-E	0.33	0.32	—	0.65
Dynamic low-E	0.45	0.53	—	0.98
Triple-pane low-E	0.71	0.31	—	1.02
High-R dynamic	1.10	0.52	—	1.62
Integrated facades	1.10	0.52	1.0	2.62

Note: Integrated facades adds dimmable daylight controls to the High-R dynamic case.

building controls, and aesthetic aspects of market acceptance. For a given building, the relative importance of each of these will vary, but collectively they help explain why significant market changes occur slowly.

We can gain some insights into the potentials for significant future change in the next 50 years by reviewing the history of technology innovation and changing market share over the past 50 years. This is explored by looking at advances in several different functional aspects of window performance, such as heat loss management, active and passive solar control, daylight savings and glare management, and integrated controls of multifunctional systems. These advances have been characterized by four interacting innovation drivers in the window industry:

1. The role of new enabling glass and coating technology,
2. The role of validated, technology-neutral metrics for rating and labeling performance,
3. The role of voluntary and mandatory standards, and
4. The role of new non-energy performance metrics driving design decisions, such as health, comfort and productivity.

3.1.1. *New glass and coating technology*

It is well known that innovation occurs slowly in the building industry, and that is reflected in the glazing and windows industries that contribute to all buildings. The glass industry had transformed the way that glass was manufactured by shifting to the "float" process, in which a continuous stream of molten glass was floated across a hot liquid tin bath, providing glass of exceptional optical quality in large sizes and a wide range of thicknesses at relatively low cost. Clear glass could be converted to bronze or grey or green to provide some solar control by adding chemicals to the inputs of the float process. Additional reflective coatings could be added for use in office towers in hot climates by evaporating a thin layer of a metal, such as gold, on one of the glass surfaces.

A snapshot of windows in the early 1970s would thus show that the windows were largely single glazed, some with storm windows, and there was some use of double-glazed windows in northern climates. In many office buildings in hot climates tinted glass was used to control solar gains and very-low-transmission reflective glass was used when the designs called for highly glazed facades.

The initial industry response to the "Energy Crisis" of the 1970s was to deploy "known" solutions to enhance energy efficiency. Given the dominance of the heating end use, this largely meant a shift to double glazing, which reduced energy loss through windows by a factor of two from single glazing but still far exceeded the losses of insulated walls. The shift from single to double glazing occurred over a 10–15-year period using incremental improvements to existing technologies, such as sealed insulating glass. But the pathway to further improvements meant going to triple glazing, which required a major redesign of the sash and frame to incorporate the wider, heavier glazing elements. However, research studies showed that adding a highly transparent, "low-emissivity" (low-E) coating to the cavity of a conventional double glazing would reduce the heat loss by another 35%, making it perform similarly to conventional triple glazing. The technical challenge, however, was to make the coating durable and highly transparent to admit daylight and useful solar energy in winter, since the initial target market was in cold, heating dominated climates. The business challenge was to make it affordable and compatible with the window industry's supply chains.

Over the course of 5–10 years a significant research investment by the US Department of Energy (DOE), in collaboration with the glass industry, resulted in the development of several first-generation low-E coatings with the required thermal and optical properties, and with a long lifetime and acceptable incremental costs.[7] Two fundamental technology innovations drove these efforts. The first, a "soft coat" approach, used sputtering, a vacuum deposition coating process that applied a multi-layer low-E coating to sheets of glass or plastic. The second, a "hard coat" process, applied a liquid or spray to the hot glass as it emerged from the float line so that the coating would be effectively fused to the hot glass surface. Sputter

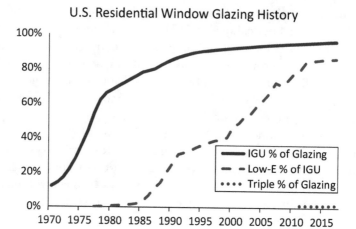

Fig. 14.   Residential window sales market share by year. Low-E shows a rapid rise in market penetration, while triple-pane market share is stagnant at around 1.4%. *Source*: Ducker.[149]

depositing the low-E coating on a thin polyester film that was then stretched across the insulated glass unit (IGU) to make a three-layer glazing was a technical success, but due to the complexities and cost of fabrication it never captured significant market share compared to similar coatings deposited directly on the glass.

These fundamental breakthroughs triggered a series of further incremental but important innovations that improved the thermal properties of existing products and expanded into new markets. Figure 14 shows the market penetration of all types of low-E coatings in the residential sector.

The data show that the low-E technology was able to saturate the market in terms of annual sales, albeit over a 30-year time frame. This broad market transformation was built on a series of intertwined technical and market advances that were mutually supportive. Since they are relevant to ongoing efforts to further deploy new technologies it is worth a quick review.

3.1.2.   *Extension of insulating value of low-E coatings*

Beginning with two pieces of glass with an air-filled cavity, the first and most important improvement is adding the low-E coating to one of the internal glass surfaces of the IGU. A second coating in the cavity has

Insulating Glazing (IGU) Solutions

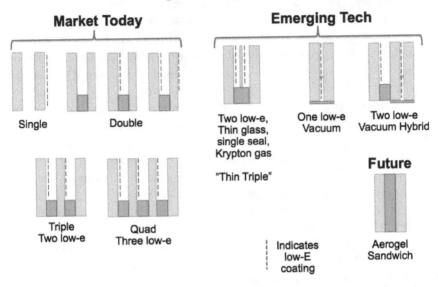

Fig. 15. Taxonomy of glazing options on the current, emerging and future markets.
Note: Low-E coated polyester film can be alternative middle glazing(s).

almost no benefit, but changing the gas fill from air to argon results in a 20% reduction in overall heat loss. Once that change is made, the next improvement is to add a low-E coating on the room-facing glass surface, which provides another 20–25% reduction in window thermal transmittance (U value). An entire taxonomy of combinations of glazing and low-E coatings and gas fills has evolved to meet different performance levels and market needs (Fig. 15).

The use of low-E coatings and gas fills provide the energy efficiency underpinnings for more than 90% of glazings on the market today. Three new "emerging" innovations in Fig. 15 that have the potential to dramatically reduce thermal losses compared to current double, low-E, gas-filled windows by 50–80% are reaching the market now. Vacuum insulating glass (VIG), which promises center glass values above R10, can be thinner and lighter than triple glazing, and is now being manufactured by several glass companies globally. A hybrid variant is also illustrated in a triple glazed format. The designs use a low-E coating and completely evacuate all air between the two glass layers; this vacuum provides even lower heat transfer rates than the use of argon or krypton gas. But the tiny spacers that keep the glass panes from touching cause some heat loss, and a perfectly hermetic edge seal is needed to hold the vacuum over decades. Prices remain high for the first generation of products but may fall with manufacturing volume and innovation. Triple glazing has never captured significant market share in the US because the heavier, wider IGUs require a redesign of the entire window. By contrast, European windows can easily accommodate conventional triple IGUs because they are traditionally wider and heavier and use tilt/turn operating panels rather than the traditional US sliding window. A second "emerging" option (Fig. 15) shows a novel "thin triple" design for triple glazing that begins with the existing double-glazed IGU, adds a very thin layer of glass in the center (~0.7 mm versus conventional 3.0 mm glass) and adds krypton gas to the narrower air spaces. Both thin glass and krypton are now commercially available in volume and at reasonable cost compared to just five years ago, providing an opportunity to develop a "drop in" replacement glazing window for almost any existing window sash/frame design. These new design options offer R8–R15 IGU properties for the next generation of "superwindows". The research challenges for these emerging highly insulating glazings are to evaluate enhanced performance, refine manufacturing process and reduce cost, and in parallel create and expand market demand.

3.1.3. *Extension of climate applicability of low-E*

Low-E coatings evolved initially as a cold climate solution since they were transparent to daylight and solar gain in climates where the extra solar gain was helpful in winter. But in southern climates that dominated construction markets for much of the last 30 years, managing cooling loads was a key performance need, while still admitting daylight in offices and permitting clear views in homes. Extensive research and development on the multilayer thin-film low-E coatings provided new "spectrally selective" versions of the coatings that transmitted visible light but reflected near infrared, thus delivering over twice as much daylight per unit cooling load as conventional glazing. These coatings have become the standard throughout the US, since even northern homes are built with air conditioning and thus need to minimize summer cooling energy use.[8] While these coatings give up some potential energy savings, the simplicity of a supply chain that standardizes on "one size fits all" has won the marketing war to date. The high-rate thin-film deposition coating systems and the manufacturing processes that produce these coatings are a tribute to innovations in materials science, plasma physics and industrial engineering — producing coatings with more than 20 layers, very high yields, very large output and costs that are remarkably low: ~$1.00/ft^2, making them affordable enough to be used in virtually all windows.

3.1.4. *Ratings and labels for market clarity and consumer investment*

Determining the performance properties of windows with potentially hundreds of types of "invisible" coatings is beyond the reach of not only homeowners but architects and other design professionals. While the proliferation of new coating and glazing options enhances long term potential savings, in the short term it can create chaos in markets if decision making is affected. With financial support from DOE and technical support from Lawrence Berkeley National Laboratory (LBNL), the National Fenestration Rating Council (NFRC) was established in 1989 as a nonprofit entity to develop an accurate, unbiased window rating system.[9] Based on a series of new validated modeling tools and certified databases, it was possible to rate and label window performance so as to provide accessible, comprehensive objective window rating data throughout the US. This information is referenced by voluntary programs such as ENERGY STAR Windows, as well as most state and national codes and standards. Numerous electric utilities, designing incentive programs to increase market share of efficient products, are able to reference both the ENERGY STAR label and the underlying NFRC ratings in support of their programs, as are non-profits established to promote window energy efficiency nationwide.[10] These labels and ratings have recently been extended to include a new class of low-E retrofit storm windows.

3.1.5. *Manufacturing innovation*

Traditional methods for developing new windows followed the classic pathways for most industrial products: build a physical prototype, test the prototype, refine the prototype based on test results, test again, etc., until performance goals are met. This is a time consuming and costly process that ultimately slows innovation. To address this, LBNL's new window rating tools[11] were not only used in ratings codes and standards but also for "rapid virtual prototyping" of new high performance windows. In a matter of hours a virtual prototype can be created and "tested"; refinements to meet performance goals can be done in hours rather than weeks. When tax credits for high performance windows were introduced suddenly in 2009, hundreds of new window designs meeting the tax credit specification were created and certified in weeks.

3.1.6. *Smart glass and automated shading*

Low-E coatings have evolved to the point where no fundamental new refinements are possible in terms of lower emittance and better light-to-solar ratios. But other critical technical challenges remain to be solved. The holy grail of high performance glass has been a dynamic, responsive glazing layer whose properties can change over time; from highly transparent at night or during cloudy weather to low transmittance on sunny days for glare control and solar load control.

There are two fundamental ways to implement strategies to change the solar control properties of a window system. One pathway to achieve that performance is traditional but is now being reinvented: adding an operable shade, blind, or other window "attachment" to the existing glazing/window system. Shades and blinds and all their variants are ubiquitous in most buildings, but they have been largely decorative in US buildings, or focused on comfort, and their performance is intermittent and uncertain since manual operation by occupants is notoriously unreliable. Most of such shading is located behind the glass inside the room for practical reasons, but exterior systems are far superior in terms of sun control capability. External versions are used extensively in Europe, where air conditioning has been less common and good thermal comfort relied on managing the solar loads at the building skin. As noted earlier there are many variants of glazings, and determining the system-level properties of all combinations of shading and glazing can be very complex. The DOE funded LBNL to extend the glass and window thermal models described above to model all combinations of glass and shading. These are now complete and available to the window industry, and are being used by a new nonprofit — the Attachment Energy Rating Council[12] — to develop energy ratings for the "attachment" world in much the same way that NFRC provides ratings for windows. The next step in extending the energy savings impact of these shading systems is to motorize and automate them so that their performance and resultant savings are maximized and can be reliably delivered. Low cost sensors, wireless controls, local PV/battery packs that eliminate the need for wiring, and more integration with other smart building controls provides the infrastructure to see rapid adoption of these approaches over the next decade.

The second, and in some ways, a more elegant pathway to provide solar and glare control on demand is to deliver these properties in the glazing itself. Active research on "smart glass" has been underway with DOE support since 1976, but only in the past 10 years have commercially viable products become available for building applications. Smart glazing solutions are available as both "passive" and "active" options in terms of the drivers that change optical properties. Passive systems change properties in response to a change in environmental conditions — a photochromic material switches to a dark state in response to ambient light levels (like common sunglasses). A thermochromic material switches in response to temperature, which means it will switch based on the combined effect of air temperature and sunlight. The passive response simplifies the operation in that no sensors, controls, or wiring is needed to power or operate the glazings. However, their response is constrained in many ways by ambient conditions, and there is no "switch on demand" capability that is of growing interest to utilities for electric load management. Active devices are based primarily on two classes of materials — electrochromic and liquid crystal devices — with most development to date based on electrochromics. These multilayer coatings can switch over a wide dynamic range (e.g., clear to dark light transmittance: 60% to 1%; solar heat gain: 0.4–0.8). Most switch at low voltages. It takes 3–30 minutes to fully switch, and they have some color associated with the dark state, typically a blue color characteristic of tungsten-based coatings. The coatings today are complex and costly, relative to low-E coatings, but promising new materials research may deliver a next generation of lower cost devices with improved properties.

The first generation of electrochromic coatings from several manufacturers tested in labs and outdoor testbeds has demonstrated compliance with durability standards over 50,000 switching cycles, and proven to deliver the expected energy savings, dynamic load response, and active glare control. However, to achieve widespread application, these solutions, like the automated shading described earlier, require a degree of systems integration with lighting, HVAC and occupant comfort that is new for the building industry. Thus, market penetration has been slow but steady. Further improvements in optical properties, lower costs, greater interest in their role with grid-integrated controls and with more seamless interoperability across controls and power supply suggest a future where 30–60% of all windows might incorporate some version of these smart glazing solutions or the related motorized shading systems.

3.1.7. *Daylighting solutions*

Daylighting has had potentially large effects on commercial buildings' energy use due to the major role of lighting in a building's energy budget. The challenge has always been twofold — to manage the daylight levels in the space as the sun and sky continually change (see the discussion above on smart glass and motorized shading for active sunlight management), and to dim the lights to save energy. The conventional rule of thumb is that a window can provide useful daylight at a floor depth up to about 1.5 times the window height, or 15 feet. But daylight at a building façade is so intense that it can in principle be used to light the building up to 40 feet from the perimeter. This doubles or triples the floorspace potentially affected for energy savings, thus enhancing the economics of savings. However, this approach requires optical systems to redirect daylight deeper in the space, without glare. While several novel optical systems are commercially available, none has achieved significant market impact to date. New systems with active tracking have better potential performance but are still in the research and development stages, and the likelihood of future commercial success is unclear even though the energy savings are potentially large.[141]

While in principle dimmable daylight-linked electric lighting controls are straightforward, in practice they have been a huge practical challenge for the industry, slowing the market impact. Numerous studies and demonstration projects over the past 30 years[13] show that when daylight solutions are properly designed, deployed, commissioned and operated, they deliver the savings that were expected. However, the bottom line track record has been mixed. Highly efficient LEDs now reduce the economic advantage of daylighting, but at the same time other factors may increase its attractiveness. The light levels of LEDs that can now be more easily dimmed to address changes in available daylight are at much lower cost than the older dimming fluorescent ballast technologies. But the real commercial sector market drivers for daylight over the next decade are unlikely to be energy related, rather, they will likely be tied to an emerging consensus about the importance of occupant health, comfort, performance, and the role that daylight and views play on those parameters. Green building rating systems are promoting performance related to greater use of high quality daylight designs. The applicability of many of these specific quantitative performance claims, for example those related to worker productivity, are open to challenge because there is not yet a way to correlate design variables such as window size or properties to these outcomes. But the overall importance of the quality of glare-free, well-daylighted spaces with views to occupants and building owners will certainly grow over time. Even though a predictive quantitative model does not yet exist that directly relates these variables, the fact that office salaries are about 100 times larger per unit floor space than the cost of energy will continue to provide a powerful incentive to further explore the role of daylight in buildings, even if the energy impacts are diminished.

3.1.8. *Power generating windows*

Many of the current generation of photovoltaic (PV) cells are built on glass substrates. Thus, it should not be surprising that there is interest in helping to reduce building energy consumption with Building Integrated PV (BIPV) by converting windows into power sources. Conventional PV is normally placed on a flat or tilted surface on the roof or ground, but there is increasing interest in incorporating them into the vertical skin of the building — both the opaque and transparent elements. Buildings in the US contain in excess of 30 billion square feet of glass that might be used, although only about one-third (Southeast-South-Southwest) is most suitable to intercept direct sunlight, and other large fractions are shaded by adjacent buildings or are otherwise unsuitable. Opaque PV panels are used as sun shades, as well as being placed on opaque building surfaces, but there is growing interest in using semi-transparent or fully transparent materials as part of the window to generate power while providing other window functions such as daylight and view. The more transparent versions have greater aesthetic value but produce only about 50% of the power of conventional cells since they absorb energy in the near IR and UV but transmit much of the visible portion of the spectrum. Many of these designs are still in development, although there are prototypes and some products in place in buildings worldwide. They have the added challenge of needing to be connected to wiring and specifically to alternating current/direct current (AC/DC) power management in buildings. There are clearly specialty applications where small amounts of intermittent power will have market value, but more development and testing is needed to assess whether this will be a major contributor to a zero net energy

(ZNE) building energy balance. The arguments for centralizing PV power relate to lower per unit costs. The arguments to use BIPV on the building relate to lower transmission losses and a potentially more resilient system in some failure modes.

3.2. Heating, ventilation and air-conditioning systems

Global demand for air conditioning (AC) is increasing continuously and significantly and is expected to triple by 2050. The global stock of air conditioners in buildings will grow to 5.6 billion by 2050, up from 1.6 billion today — which amounts to 10 new HVAC systems sold every second for the next 30 years, according to the IEA report *The Future of Cooling*.[14] HVAC systems consumed 41% of the primary energy use in US buildings in 2010. Reducing HVAC energy use is a critical path to low-energy buildings. HVAC technologies include high-efficient equipment, system-efficient design, advanced controls and passive cooling technologies that have the potential to reduce HVAC energy use by 50%. Emerging HVAC technologies usually can achieve direct benefits of energy savings, demand reduction and co-benefits of improving occupant thermal comfort, indoor air quality (IAQ), and thus occupant productivity, health and well-being. Other co-benefits include energy flexibility, i.e., the capability to adjust HVAC operation in response to demand response (DR) and renewable energy availability, which is a key feature of grid-interactive efficient buildings.

To reduce HVAC energy use, high-efficient equipment such as rooftop units (RTU), chillers, boilers, fans and pumps is important but not sufficient. Building designers should adopt system-level performance approaches that integrate all components and consider their connections and interactions, to ensure that all system components operate in tandem to achieve cohesive high efficiency. Advanced controls play a decisive role in HVAC efficient operation. Traditional full-time-full-space HVAC control, i.e., conditioning all the spaces all the time, wastes lots of energy. Part-time-part-space HVAC control, i.e., conditioning only the occupied spaces during the occupied times, is becoming a new trend.

This section describes several currently in-use and emerging HVAC technologies that have high potential for energy savings and occupant comfort improvements. However, these options are not intended to be exclusive; the best HVAC technologies are always building-specific and depend on building characteristics, operation strategies and the building occupants' behaviors.

3.2.1. *Variable refrigerant flow systems*

A VRF system is a refrigerant system that varies the refrigerant flow rate using variable speed compressor(s) in the outdoor unit and electronic expansion valves (EEVs) in each indoor unit. The system meets the space cooling or heating load requirements by maintaining the zone air temperature at the setpoint.[15] Its ability to control the refrigerant mass flow rate according to the cooling and/or heating load enables the use of as many as 60 or more indoor units with differing capacities in conjunction with one single outdoor unit. This unlocks the possibility of having individualized comfort control, simultaneous heating and cooling in different zones, and heat recovery from one zone to another.

VRF systems have either two or three-pipe configurations. The two-pipe VRF system, known as the VRF Heat Pump system, is the most general type that can be used for cooling or heating, but it cannot be used for both simultaneously. The three-pipe (a high pressure gas pipe, a low pressure gas pipe and a low pressure liquid pipe) VRF system, known as the VRF Heat Recovery system, can deliver simultaneous heating and cooling to different zones by transferring heat between the cooling and heating indoor units, as shown in Fig. 16. Simultaneous heating and

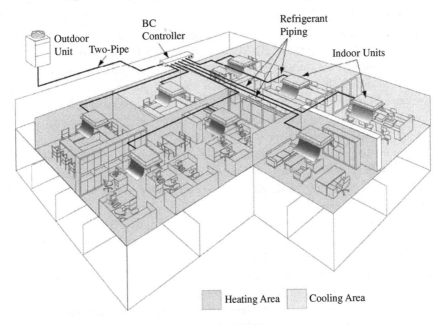

Fig. 16.   Example layout of a VRF heat recovery system.

cooling generally occurs in the winter season in medium-sized to large-sized commercial buildings with a substantial core area or computer rooms.[16]

VRF systems can provide flexible controls, better thermal comfort and consume less energy. This is because of multiple advantages of the VRF systems, including: (1) more efficient operation during part load conditions with the help of variable speed compressor and fans, (2) minimal or no ductwork reduces air leakage and heat losses, and (3) smaller indoor unit fans that consume less energy while reducing indoor noise.[15,17–19] A typical VRF system has one outdoor unit serving multiple indoor units. Each indoor unit can have its own thermostat to control its operation, meaning it can be turned off if the zone is not occupied or the thermostat can be adjusted to meet the occupants' thermal comfort needs. The individual and flexible control capabilities make VRF systems a perfect match for applications that require personalized comfort conditioning. As a result, VRF systems are becoming more widely used, with sales booming worldwide.[142] Hospitals and nursing homes are good candidates for the VRF system, since they avoid zone-to-zone air mixing. Residential buildings, especially luxury single-family homes, condos and multifamily residential buildings, also tend to use VRF systems.[20] Also, their quiet operation enables VRF systems to be installed in buildings where strict noise regulations apply, e.g., school buildings.[21]

As one of the emerging HVAC technologies, VRF systems have been comprehensively compared with conventional air conditioning systems, such as variable air volume systems, fan coil systems and package ducted systems. These comparisons found that VRF systems consume less energy than conventional air conditioning systems: 20–58% less than variable air volume systems in the cooling season;[20,22–24] 10% less than fan coil plus dedicated outdoor air systems in the cooling season;[22] 35% less than central chiller/boiler systems under the humid subtropical climate conditions;[16] and 30% less than chiller systems under the tropical climate conditions.[25]

It should be noted that actual savings from VRF systems vary depending on several factors, including climate, operation conditions and control strategies. The flexibility of zoning and control collectively contribute to extra potential energy savings for buildings, especially those with diversified zonal loads, such as residences.[26–28]

From the perspective of thermal comfort, the individual control allowed by the VRF system enables users to adjust the thermostat settings according to their specific requirements, hence it improves thermal satisfaction.[29,30] This was proved by a field-performance test of two different control modes (individual and master) that were applied to a VRF system in a test building.[20]

Therefore, the VRF system not only consumes less energy than the common air conditioning systems, it also provides better indoor thermal comfort due to its independent zoning controls. VRF systems can be 10 to 50% more expensive than traditional systems, which is one of their main drawbacks, but the significant energy saving potential can support a reasonable payback period,[16,25] in addition to potentially improved productivity from better occupant comfort and space saving from reduced or no ductwork.

3.2.2. *High energy efficiency rooftop unit*

In the US, packaged equipment such as RTUs are the most common type used for commercial space cooling, serving 38% of the floor space overall. There is significant energy saving potential from improving RTU energy efficiency, which is typically represented by the energy efficiency ratio (EER) for rated condition and the integrated energy efficiency ratio (IEER) for weighted load conditions. Inefficient RTUs not only waste significant amounts of energy, but also are a key strain on the regional grid during peak demand hours, which is an increasing concern for utilities, policymakers and energy regulators.[31]

The DOE's Building Technologies Office initiated a High Performance Rooftop Unit Challenge to "urge manufacturers to build and deliver innovative, competitively priced, energy-saving rooftop units" that, among other specifications, must achieve an IEER of 18.0 or more. Multiple manufacturers such as Daikin, Carrier, Lennox and others have participated and met the challenge. Products that meet the challenge specifications are able to boost the IEER to as much as 50% greater than the EER by focusing on part-load performance improvement (while still increasing EER to greater than 13).[32] These high performance RTUs can dramatically reduce energy consumption and reduce peak electric demand.

3.2.3. *Dedicated outdoor air system*

A DOAS is a 100% outdoor air system providing ventilation air for occupants in buildings. Usually one or more separate HVAC systems provide space cooling and/or heating. Conventional HVAC systems mix ventilation air and recirculation air to provide both ventilation and cooling and heating. In comparison, system design with DOAS provides multiple benefits: (1) easy ventilation control to meet occupants' dynamic demands, (2) easy implementation and operation of heat recovery between the outdoor air and relief/exhaust air in cold or hot climates, and (3) decoupling of sensible and latent cooling, as most moisture in buildings is in the ventilation air, especially in humid climates. Figure 17 compares a conventional VAV system with a DOAS coupled with a VAV system.

Various DOAS designs differ in the way outdoor air is delivered to space or occupants. The DOAS can deliver outdoor air directly to spaces, to the inlets of zone terminal equipment, or to the outdoor air inlets of air handling units. The sizing of the DOAS should be based on the maximum outdoor air determined by the design occupancy defined in ASHRAE Standard 62. However, because occupancy varies and demand is much lower than the design's full occupancy rate most of the time, the DOAS should have variable speed fans that enable users to adjust the outdoor air flow. DOAS control should enable an economizer mode, that is, maximizing the use of outdoor air when it can be used to provide free cooling.

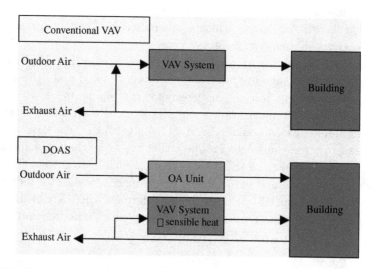

Fig. 17. Comparison between a conventional VAV system and a DOAS coupled with a VAV system.

Source: UPONOR.[150]

Unusual events such as wildfires may cause a sudden increase in outdoor air pollution (e.g., an increase of fine particulate matter [$PM_{2.5}$] or ozone). To maintain good indoor air quality, the DOAS should be able to react to such events by minimizing the use of outdoor air or shutting off all air flow from exterior sources.

Optimal design of the DOAS needs to consider many factors, including the design supply air temperature — whether to provide neutral air at room temperature or cooler air to the space. The latter helps remove moisture from the outdoor air and handles a portion of the space cooling loads. HVAC manufacturers nowadays provide various configurations of DOAS to meet customers' diverse needs.

3.2.4. *Radiant heating and cooling*

A radiant heating and cooling system refers to temperature-controlled surfaces that exchange heat with their surrounding environment through convection and radiation. By definition, in radiant heating and cooling systems, thermal radiation covers more than 50% of heat exchange within the space.[33] Radiant heating and cooling systems use panels or components embedded in floors, ceilings or walls. Hydronic radiant systems are the most common, but other available types include air-based and electrical systems.

Radiant heating has a number of advantages. It is more efficient than baseboard heating and usually more efficient than forced-air heating because it eliminates duct losses. People with allergies often prefer radiant heat because it does not distribute allergens like forced air systems can. Hydronic (liquid-based) systems use little electricity, a benefit for homes off the power grid or in areas with high electricity prices. Hydronic systems can use a wide variety of energy sources to heat the liquid, including standard gas- or oil-fired boilers, wood-fired boilers, solar water heaters, or a combination of these sources.[34] Efficient radiant heating systems are promising technologies for energy saving in commercial buildings and improving occupant thermal comfort.[35]

Radiant cooling, as shown in Fig. 18, can potentially save energy as well as achieve better thermal comfort.[36] However, a radiant cooling system may cause condensation on the cooling surfaces under hot and humid conditions during the cooling season, which is a major concern for adopting radiant cooling systems. This issue may be addressed by integrating radiant cooling with an auxiliary air-conditioning system for

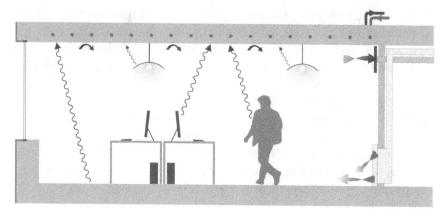

Fig. 18. Diagram of a room cooled with a high mass radiant ceiling slab. The slab is absorbing heat radiated from the people, surfaces, equipment and lights in the room. *Source*: Wikipedia, Radiant heating and cooling.

dehumidification purpose.[37] However, this will add further to the capital cost.[38] In buildings with existing hydronic radiant heating systems, it could be particularly economical to take advantage of the same principle using chilled water for cooling.

Integrated heat pump (IHP) system for space and water heating.

An IHP system can provide space conditioning while maintaining comfort and meeting domestic water heating needs. The concept is to merge several end uses together into a new solution that benefits from energy cascading, where the waste (or residual) heat from one process provides the energy input for another, as illustrated in Fig. 19. For example waste heat from AC can be used to heat water. Available IHPs include air-source, ground-source and fuel-fired IHP systems.

Previous research examined the energy saving potentials of an IHP over a conventional heat pump with electric resistance water heating. The results indicate that an IHP can produce an average saving of 30% over the baseline system with the same energy efficiency.[39] With significant energy saving potentials, the IHP systems are targeted to be part of the solution in the development of zero-energy homes.

For the past ten years, the DOE has been collaborating with manufacturers to develop a new generation of air-source IHP (AS-IHP). The concept consists of a high-efficiency air-source heat pump (ASHP) for space heating and cooling services, and a separate heat pump water

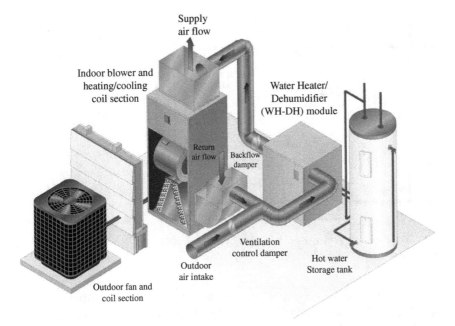

Fig. 19.   AS-IHP system concept sketch.
Source: Oak Ridge National Laboratory.

heater/dehumidifier (WH/DH) module for domestic water heating and dehumidification (DH) services. A key feature of this system approach, with the separate WH/DH, is the capability to pretreat (i.e., dehumidify) ventilation air and dedicated whole-house DH independent of the ASHP. Based on the demonstrated field performance of the AS-IHP prototype and estimated performance of a baseline system operating under the same loads and weather conditions, a bin analysis estimated that the prototype would achieve ~30% energy savings relative to the minimum efficiency suite.[40]

3.2.5. *Personalized conditioning system*

Occupants have diverse needs of thermal comfort and indoor air quality. Their preferred temperature for cooling and heating vary, and the desire for more or less air movement and outdoor air varies by individual. In addition, it wastes energy to condition the entire space when only some portion of it is occupied. The concept of a personalized conditioning

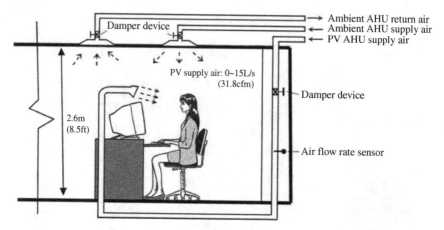

Fig. 20. Example of a personalized conditioning system (AHU refers to air handling unit).[43]

system (PCS) has emerged as an approach to provide individual comfort and to reduce energy use (Fig. 20). One PCS design uses a combination of ambient conditioning and workstation conditioning. The former provides a neutral thermal environment for the entire space while the latter enables individuals to adjust supply air volume or temperature to meet their varying comfort needs. The "Berkeley Chair", developed by the Center for the Built Environment of the University of California, Berkeley, provides local cooling and heating using components (fans and electrical heaters) built into the chair. Comfy[41] is another type of PCS; it provides an app for occupants to request a short-time air jet to quickly cool them down if they feel warm or hot. Emerging PCS technologies may also use occupant sensing and tracking to detect occupants and memorize their thermal comfort preference and feed the information to HVAC control systems. Studies[42–44] have shown that PCS technologies have significant energy savings potential, as well as improving an occupant's level of comfort and productivity.

3.2.6. *Low global warming potential (GWP) refrigerant*

Traditional vapor-compression heat pumps and air conditioners reply on refrigerants as working fluids. However, refrigerants commonly used today, such as hydrofluorocarbons (HFC), have a significantly higher GWP than carbon dioxide when they are released to the atmosphere. For example, the refrigerants R-410A and R-134a have GWP factors of 1,700 and 1,430, respectively. It is important to develop alternative refrigerants that not only reduce environmental impacts but also enhance/maintain performance.

Most countries already are implementing policies that call for the phasedown of the production and usage of HFCs. For example, the European Union's F-gas regulation is set to reduce the availability of HFCs by 79% by 2030. In addition, most countries are signatories to the Kigali Amendment that also calls for the phasedown of HFCs in the future. Even in the US, which is not currently a signatory to the Kigali Amendment, several states, including California, are moving aggressively to phase down the use of HFCs.[45]

The development of low-GWP refrigerants is ongoing, and promising new composites are announced frequently. Currently, several alternatives are in use on a limited basis:

- **HC (Hydrocarbons):** This group consists of refrigerants like isobutane (R600a) and propane (R600a). These materials score well for GWP, toxicity, pressure and availability, but they have flammability issues.
- **Carbon dioxide:** Carbon dioxide refrigerants currently include R744, which has an extremely low GWP but operates at a peak pressure of 1,740 pounds per square inch (psi), which is significantly higher than the current average peak pressure of 290 psi in HFCs.[46]
- **Ammonia:** Ammonia refrigerant is available as R717. It has a low GWP, however, it is toxic and flammable.
- **R32:** R32 scores well on toxicity but it is flammable, and availability is currently limited.

In addition to the refrigerants mentioned above, the industry is working on stable refrigerant alternatives, including unsaturated hydrochlorofluorocarbon (HCFO) and hydrofluoroolefin (HFO, e.g., 1234yf). Table 4 compares the properties of available low-GWP refrigerant options. These leading-edge refrigerants are designed to substantially lower flammability or eliminate it altogether.[47]

Electrochemical compression systems are also an emerging technology that may utilize low-GWP refrigerants. As high-GWP refrigerants phase out of the HVAC industry, electrochemical compressors may have more attractive economics and efficiency compared to new vapor-compression systems.[32]

Table 4.　Respective properties of the low-GWP refrigerants under consideration.

Refrigerant	HFC	HCs	Ammonia	CO$_2$	HFO 1234yf
		Natural			HFO
GWP (100 years)	✗✗	✓	✓✓	✓✓	✓
	R134a 1300–R410A 1900	3–5	0	1	4
Toxicity	✓✓	✓✓	✗✗	✓	✓✓
Flammability	✓✓	✗✗	✗	✓✓	✗
Materials	✓	✓	✗	✓	✓
Pressure	✓	✓	✓	✗✗[a]	✓
Availability	✓✓	✓	✓	✓	✗✗
Familiarity	✓✓	✓	✓	✗	✗

Notes: Very poor ✗✗ Poor ✗ Good ✓ Very Good ✓✓
[a]CO$_2$ has been categorized "very poor" in terms of pressure because the RAC industry will need to learn to cope with using a fluid at 120 bar, which is much higher than the current peak pressures of around 20 bar. However, the high pressure does deliver some desirable characteristics such as smaller pipe diameters and less compressor swept volume.
Sources: F-gas support Information Sheet RAC7 alternatives; Brown *et al.*[46]

3.2.7.　*Other technologies*

Other HVAC technologies can, if applied correctly, save a significant amount of energy and/or energy cost in buildings. These include solid-state cooling (thermoelectric), energy storage using phase change materials, heat pumps that can operate at low ambient temperature, DC-driven motors and equipment, combined heat and power systems to provide both heat and electricity, and integrated solar thermal and power systems. It should be noted that energy storage in building shells or phase change materials (including ice) is far cheaper than electric storage technologies. The energy storage materials could play a key role in facilitating the introduction of intermittent renewables and otherwise managing "smart grids" that achieve systemwide efficiency gains.

3.3.　*Lighting*

The rapid emergence of light-emitting diodes (LEDs) has been transformative for the lighting industry. For more than 100 years, this industry had been relatively static, with evolutionary technological developments slowly bringing new light sources and equipment to markets. Different light sources dominated different applications based on their particular advantages: low initial cost and good color helped incandescent lamps take over the residential market, lower lifecycle costs led to fluorescent dominance in commercial interiors, and high efficacy and high output made high-intensity discharge (HID) lamps a good fit for industrial and exterior applications. Today, the "LED revolution" is nearly complete, as current LEDs are superior to these legacy light sources in almost every way. LEDs are capturing the last niches that had previously eluded them because of cost, color or output limitations. As a result, the installed building stock of light sources is projected to continue to rapidly shift toward LEDs, as seen in Fig. 21.

LEDs are solid-state devices that convert electrical energy into light. Some LEDs convert electricity directly to visible light but increasingly today's LEDs utilize a phosphor coating on top of an LED chip. In these applications, the LED chip typically generates blue and/or UV light while the phosphor coating converts the light output to a more full-spectrum, white light. Based on the LED chips and phosphors used, manufacturers can tailor the light output and color characteristics of the LED package to their desired specifications. LED chips are combined with substrate boards and encapsulants to form an LED package. The LED packages are then combined with drivers, heat sinks, secondary optical systems and other components to make LED lamps and LED luminaires (Fig. 22).

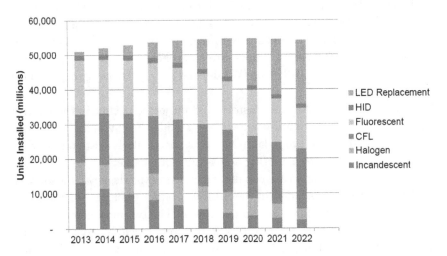

Fig. 21. Number of light sources in the US building stock by lighting technology. *Source*: US DOE, 2016.

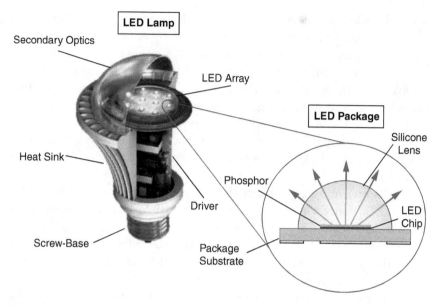

Fig. 22.   Anatomy of an LED package and an LED lamp.
Source: US DOE, 2016.

The market dominance of LEDs of the last decades is closely tied
to their rapidly increasing efficacy (measured in lumens [lm] per watt)
and decreasing cost. Figure 23 shows the past and projected efficacy
of LEDs between 2005 and 2025. LEDs have increased in efficacy from
50 lm/W in 2005 to more than 150 lm/W today, and are projected
to approach 250 lm/W by 2025. Efficacy improvements are expected
to slow as the theoretical limits are approached — these theoretical
limits are between 250–350 lm/W, depending on the color of the light
produced.

Figure 24 shows the rapid historic and projected decline in the relative
cost of 800 lm A-lamps (also known as "standard light bulbs") between
2013 and 2020. While the A-lamp market is particularly cost-competitive,
similar price drops are occurring across all LED markets.

LEDs are inherently easier to dim than most other legacy light sources,
and this feature is included in most LED product offerings. Increasingly,
some LED products offer color tuning as well. The two most common types
of color tuning are full color control and "tunable white" control. In full
color control, red, blue and green (RGB) LEDs are each independently
controlled so they can be combined to provide nearly any desired color. In

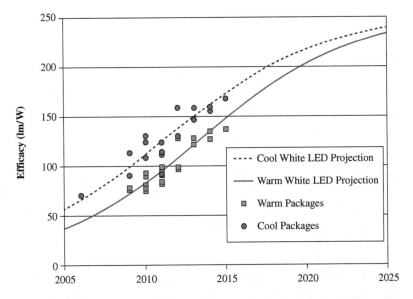

Fig. 23. Historic and projected efficacy of warm color and cool color LED packages.
Source: US DOE, 2016.

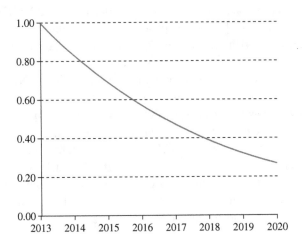

Fig. 24. Relative price of 800 lm A-lamps between 2013 and 2020.
Source: US DOE, 2016.

tunable white applications, two or more LEDs with different color characteristics (e.g., "warm white" and "cool white") are typically combined to
provide white light at the desired color temperature.

The following LED definitions provide a more detailed overview of LED technologies (*Source*: ENERGY STAR):

- **Light-emitting diode (LED):** A p–n junction solid-state device of which the radiated output, either in the infrared region, the visible region or the ultraviolet region, is a function of the physical construction, material used and exciting current of the device (10 CFR 430.2).
- **LED array or module:** An assembly of LED packages (components) or dies on a printed circuit board or substrate, possibly with optical elements and additional thermal, mechanical and electrical interfaces that are intended to connect to the load side of a LED driver. Power source and ANSI standard base are not incorporated into the device. The device cannot be connected directly to the branch circuit (ANSI/IES RP-16-10).
- **LED light engine:** An integrated assembly composed of LED packages (components) or LED arrays (modules), as well as an LED driver and other optical, thermal, mechanical and electrical components. The device is intended to connect directly to the branch circuit through a custom connector compatible with the LED luminaire for which it was designed. It does not use an ANSI standard base (ANSI/IES RP-16-17). For purposes of this specification, light engines that rely on the luminaire for optical control and/or thermal management, assemblies featuring remote-mounted drivers ("non-integrated"), and/or GU24 based integrated SSL sources not in the scope of the ENERGY STAR Lamps specification shall also be considered LED light engines.
- **LED luminaire:** A complete lighting unit consisting of LED-based light emitting elements and a matched driver together with parts to distribute light, to position and protect the light emitting elements, and to connect the unit to a branch circuit. The LED-based light emitting elements may take the form of LED packages (components), LED arrays (modules), an LED Light Engine, or LED lamps. The LED luminaire is intended to connect directly to a branch circuit (ANSI/IES RP16-17).
- **LED package:** An assembly of one or more LED dies that includes wire bond or other type of electrical connections, possibly with an optical element and thermal, mechanical and electrical interfaces. Power source and ANSI standardized base are not incorporated into the device. The device cannot be connected directly to the branch circuit (ANSI/IES RP-16-17).

- **Solid-State Lighting (SSL):** The term "solid-state" refers to the fact that light is emitted from a material by a semiconducting process of electron transition from a conduction band to a valence band whether or not the wavelength of this light is converted by additional components.

3.3.1. *Lighting controls*

The term *lighting controls* broadly refers to technologies that adjust the output of lamps or luminaires. Lighting control technologies include manual controls, occupancy sensors, daylight harvesting, dimming systems and schedulers. The following factors have stimulated the market for lighting controls systems in recent years:

- **LED controllability:** A major driver is the relative ease with which LEDs can be controlled. Whereas dimming ballasts for fluorescent lamps always represented a significant incremental cost increase over standard (on/off) ballasts, dimming drivers for LEDs can be added for little to no cost over on/off drivers. This of course enables dimming systems but also other controls system that benefit from inexpensive, continuous dimming, such as daylight harvesting systems.
- **Wireless technologies:** Wireless technologies and protocols, including Wifi, Bluetooth and Zigbee, have greatly increased the reach and decreased the costs associated with controlling light sources. Many lamps and luminaires are now "smart" systems that have wireless radios built in, allowing them to be individually controlled, rather than controlled in larger banks, as lights were traditionally. Similarly, occupancy sensors, daylight sensors and manual switches can now be easily placed in appropriate locations without considerations to wiring limitations.
- **LED efficacy:** Another driver relates to decreased lighting loads due to the increasing efficacy of LED systems. On one hand, this reduces the energy savings associated with lighting control systems because they are controlling smaller loads. But ironically this has led to an increased demand for lighting controls as energy efficiency standards and programs have been more aggressive in looking for lighting energy savings and increasingly looking to lighting controls systems to provide those savings. Once LEDs have been installed, there are few opportunities to increase the efficacy of the light sources, so the only remaining

energy savings opportunities are those offered by lighting controls systems, which can reduce the operating hours and/or light and power levels of the lights that are installed.

3.3.2. *Light and health*

Human biology has evolved to have two distinct optical systems: (1) the visual system, by which we see and process images, and (2) the circadian system, which regulates our biological clock and associated biological systems (Fig. 25). These two systems have significantly different spectral and temporal responses to optical input. Specifically, the visual system peaks at 555 nm (green) and responds nearly instantaneously to inputs, while circadian stimulation peaks at 460 nm (blue/violet) and responds after several minutes of optical activation. Exposure to too little light during the day and/or too much light in the evening can lead to circadian disruption, which has been linked to poor sleep and performance and increased risk for diseases, such as diabetes and cancer.

Historically, all the lighting systems designed and installed in buildings considered only the photopic (visual) system, and all light meters used to characterize lighting in buildings were calibrated to measure photopic light. This is starting to slowly change as the research community develops a deeper understanding of the impacts of light on health, and the lighting industry develops new technologies and projects that leverage this work.

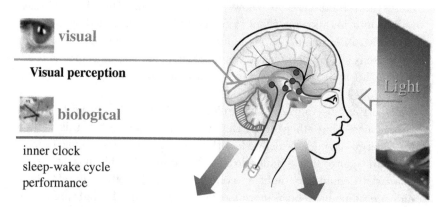

Fig. 25. Humans have two distinct optical systems: visual (photopic) and biological (circadian).

Source: US DOE, 2016.[48]

The most notable lighting technologies that exist today that focus on light and health are dynamic color tuning systems, also known as "circadian-effective lighting". These systems attempt to mimic the spectrum and intensity provided by daylight by providing high levels of cooler (blue) light in the mornings and lower levels of warmer (red) light later in the day and into the evening. Like daylight, these systems support occupant circadian systems suppressing melatonin production in the morning and promoting melatonin production in the evening.

3.4. *Miscellaneous electrical loads*

Miscellaneous electrical loads (MELs) consist of a wide range of devices that do not clearly fit in the conventional categories or end uses. They are difficult to characterize because membership in this group is often defined by exception, for example, "not space heating, not lighting, etc." rather than the services that they deliver.[49] A typical device in the MELs category consumes about 10 terawatt-hours (TWh)/year nationwide. While this is a considerable amount of energy, it is still below the threshold for most efficiency regulations. But since MELs are responsible for about 30% of electricity use in US buildings,[50] this group of devices warrants attention by building designers, policymakers and consumers. It is also the most dynamic sector, where new devices appear and vanish much faster than the standard end uses.

3.4.1. *What does the MELs category include?*

The MELs category includes hundreds of devices in residential and commercial buildings (and to a lesser extent, industrial buildings). Inside the residential sector, almost half of the total electricity use in 2015 was attributed to "other" uses beyond space heating and cooling, water heating, refrigeration and lighting.[51] The EIA estimated contributions of important MELs devices, including TVs, microwave ovens and pool pumps (Fig. 26). Nevertheless, 13% of the total residential electricity end uses were still not classified.

Note that although many devices provide HVAC-related services (such as humidifiers or ceiling fans), they are categorized as MELs because they are plugged into standard outlets. Still not listed are the growing number of kitchen devices, home medical devices, and aquariums, terrariums and other pet-related devices. Battery-operated mobile equipment,

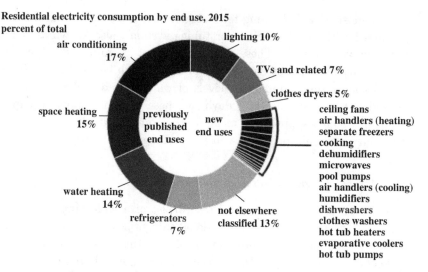

Fig. 26. Residential electricity consumption by end use, 2015.
Source: EIA, 2018b.[147]

from plug-in lawn mowers, electric vehicles and scooters to vacuum cleaners, will almost certainly be a significant fraction of residential electric use before 2025.

Data for commercial buildings is even more scarce.[48] Examples of MELs devices include computers, printers, IT infrastructure (routers, switchers, etc.), uncontrolled HVAC equipment (such as hot water recirculation pumps), emergency lighting and elevators. High voltage transformers inside commercial buildings are also sometimes considered part of the miscellaneous category.

Even though MELs defy precise definition, several cross-cutting features are apparent. Within the population of MELs, one sees the following:

- An increasing share of electronic devices.
- Devices that enable the disintegration of traditional end uses, such as microwave ovens and countertop appliances replacing stoves and ovens.
- A large fraction of energy consumed while in standby modes.
- The internet of things (IoT) revolution, with continuous network connectivity.
- Devices that reflect trends or changes in culture or demographics.

Some devices draw large amounts of electricity but are present in relatively few buildings. Elevators are an example. Other devices consume small amounts of electricity but are ubiquitous. The Ground Fault Circuit Interrupters (GFCIs), which are required in all outlets near water, draw only about 1.5 W each. However, a billion of them are already installed in buildings.[51] These features work together to make the population of MELs dynamic and somewhat unpredictable.

3.4.2. *Opportunities for saving energy*

The opportunities to save energy in MELs depend on which devices are being examined, but they are often surprisingly large. As for any device, the three standard energy-saving approaches — reducing load, improving conversion efficiency and reducing standby consumption — still apply. The most efficient MELs sometimes use less than half the energy of what their competitors use, so savings opportunities clearly exist without technical innovations. For the ubiquitous devices that consume relatively little energy in the first place, the economics of making these types of improvements are challenging. Improvements will reduce each device's electricity use by as little as 1 W; that is, 9 kWh/year, which means the transaction costs may exceed the value of the energy savings. There has been interest in reducing energy use of the high-consuming MELs, notably high-voltage transformers and elevators, but many other, less prominent MELs have been ignored. In both situations, governments often do not regulate the products' efficiency because the national savings potential does not exceed the minimum threshold. Medical and safety devices are expressly exempted from minimum efficiency regulations.

In spite of their diversity, MELs could incorporate cross-cutting (or "horizontal") technologies to save energy economically. These measures include the following:

- More effective and deeper "sleep" modes after periods of inactivity,
- Low or zero standby power use, relying on technologies like energy harvesting and wake-up radio,[52,53]
- The ability to receive network signals to enter "away" modes when external information indicates that the device's service is not required, and
- Self-measurement and reporting energy consumption to building occupants or the building management system.

Many of these solutions rely on technical standards and specifications that ensure interoperability and common communications protocols. To date, however, competing protocols and technical standards have prevented the full potential energy savings from being realized. This suggests a different role for policymakers in dealing with the MELs sector.

3.5. *Building performance simulation tools to assist building design and operation*

Building performance simulation (BPS) has played a growing role in the design and operation of low-energy, high performance buildings and development of policies that drive the achievement of the US federal and states' building energy goals. *Building performance simulation* is defined as the use of computational mathematical models to represent the physical characteristics, expected or actual operation, and control strategies of a building (or buildings) and its (their) energy systems. BPS calculations include building energy flows, air flows, energy use, thermal comfort and other indoor environmental quality indexes such as glare.

From a practical perspective, BPS is commonly used to: (1) perform load calculations in support of HVAC equipment selection and sizing, (2) demonstrate code compliance of a building by comparing the energy performance of the proposed design with the code baseline, and (3) evaluate and compare design scenarios. Note that although the use of BPS in the building design process is widespread, its use in the operation, control and retrofit of existing buildings remains limited.

The *Building Energy Software Tools Directory*[143] lists a wide range of BPS tools, which can be used for various purposes by architects, engineers and other stakeholders. It is important to select a BPS tool that is appropriate for the particular application of interest — indeed, this notion is the basis of the fit-for-purpose modeling concept.[54] Stretching a BPS tool beyond its intended scope of use should be avoided; this practice may lead to modeling errors, and at minimum its use requires a deep understanding of the BPS program in question.

Adequate efforts to collect supporting data for BPS are crucial to avoid the "garbage in, garbage out" aphorism. When modeling new buildings, for example, users must anticipate how the building will be used and accurately specify design performance goals. When modeling existing buildings, on-site inspections and energy audits can be used to establish reliable input data for energy models. Sound data collection is not replaced by parallel efforts, e.g., model calibration that attempts to fine-tune key model input parameters. Data for many parameters are needed to build detailed energy models using BPS tools. Therefore user experience is needed to focus data collection around the most important model input parameters.

3.5.1. *Simulation tools use in building design and operation*

Building energy use is influenced by six factors[54]: building envelope, building equipment and system, weather, indoor environmental criteria, operation and maintenance, and occupant behavior. When simulation tools are used to calculate thermal loads or energy use in buildings, these factors should be represented at a level of detail that is adequate for the use case. To evaluate zonal HVAC systems (e.g., VRF systems) and controls, it is useful to have detailed zoning descriptions in energy models. Detailed occupant schedules from measurements or simulated by tools like the Occupancy Simulator[55] at the zone level should be used to evaluate occupancy-driven controls (e.g., demand-controlled ventilation [DCV], lighting and HVAC controls). As multi-year weather data become available designers can run simulations using multi-year AMY (actual meteorological year) weather data to understand the variations of building performance, which can guide decision making.

Building simulations are popularly used in the building design phase to calculate loads and size, select HVAC systems, demonstrate code compliance, and evaluate design alternatives aiming to achieve certain energy targets, such as LEED buildings, or the Leadership in Energy and Environmental Design from the US Green Buildings Council. Energy models are created by inputting the building and system characteristics, as well as the design basis, either manually or preferably by importing them from building information models such as Green Buildings XML (gbXML) or Industrial Foundation Classes (IFCs) from the International Alliance for Interoperability, that can be generated from computer-aided design tools used by architects or designers.

For existing buildings, detailed energy models created with BPS programs can be used to explore and evaluate energy conservation measures (ECMs) for energy retrofit projects. Usually, the base case model is calibrated using monthly utility bill data before being used to simulate and analyze ECMs.

Most buildings do not perform as well in practice as their energy performance levels deteriorate over time. Reasons for this deterioration in performance include faulty construction, malfunctioning equipment, incorrectly configured control systems, and inappropriate operating procedures. One approach to addressing this problem is to compare the predictions of an energy simulation model of the building to the measured performance, and analyze significant differences to infer the presence and location of faults. Model-based retro-commissioning refers to this use of building energy models to help identify and evaluate operation problems in buildings as part of a retro-commissioning process. Calibrated energy models can be a good tool for identifying malfunctioning equipment and in assisting the measurement and verification (M&V) of a retro-commissioning project. As an example, Marmaras[56] discussed how building energy models can be used in the retro-commissioning process of an underperforming LEED Gold-level certified police station.

Building control systems are critical to ensuring efficient operations and occupant comfort. To support building control, BPS is being coupled in real time with building energy monitoring and control systems and sensors, where it is used to predict thermal loads in buildings and provide guidance on energy and comfort-optimal control strategies (e.g., set point adjustments, charging and discharging of energy storage, DR strategies). Real-time building operation data (equipment and systems), predictive weather data and occupant data can feed data to energy models that simulate and evaluate various control strategies across a future time horizon, identifying the control strategy with the best predicted energy and comfort outcomes. This type of model predictive control is an advanced method of process control that has been in use in chemical plants and oil refineries since the 1980s, and is only recently being appropriated for power system balancing models and building controls in large-sized commercial buildings.[57,58,59]

3.5.2. *Building performance simulation tools*

There are various building performance simulation tools used by practitioners and researchers to support the aforementioned use cases. A few popular tools used worldwide include: EnergyPlus,[60] DesignBuilder, DeST[61] ESP-r,[62] IDA-ICE,[63] TRNSYS[64] and IES-VE. Table 5[151] illustrates the characteristics of some prevailing simulation tools.

EnergyPlus is the DOE flagship building energy simulation tool. It builds upon physics-based surface and zone heat and mass balances, and calculates the energy flows in buildings, as well as on-site renewable power

Table 5. Characteristics of reviewed simulation tools for building energy demand.

	Transient/ static	Radiation analysis	Coupling with GIS	Building geometry	Energy demand modelling approach	Energy demand modelling	Types of building energy demand	Impact of user behavior on energy demand	Modelling of energy generation	District thermal network	Electricity network	Gas network	Building design optimization	Time scale	Availability
DOE-2	T	x	—	x	Eng	Endo	H, C, V, L	—	PV, GSHP, TES, boiler, chiller	—	—	—	x[a]	Hourly	Free
eQUEST	T	x	x	x	Eng	Endo	H, C, V, L	—	PV, CSHP, TES, boilers, chillers	—	—	—	x[a]	Hourly	Free
ESP-r	T	x	—	x	Eng	Endo	H, C, A, L, DHW	x	PV, solar thermal, boilers, chillers	—	—	—	x[a]	Sub-hourly	Free, open-source
Energy Plus	T	x	x	x	Eng	Endo	H, C, A, L, DHW	x	PV, BIPV, cogeneration, boilers, chillers	—	—	—	x[a]	Sub-hourly	Free, open-source
HASP/ ACLD	T	—	—	x	Eng	Endo	H, C	—	—	—	—	—	—	Hourly	Free
HOT2000	T	—	—	x	Eng	Endo	H, C	—	—	—	—	—	—	Annual	Free
TRNSYS	T	x	x	x	Eng	Endo	H, C, V	x	PV, solar thermal, GSHP, TES, boilers, chillers	x	—	—	x[a]	Sub-hourly	Commercial, open-source
Modelica	T	x	—	x	Eng	Endo	H, C, A, L	x	PV, GSHP, boilers, chillers	x	x	—	x	Sub-hourly	Open-source, free, commercial
Polysun	T	x	—	x	Eng	Endo	H, C, E, DHW	x	PV, solar thermal, wind, GSHP	x	x	—	x[b]	Sub-hourly	Commercial
IDA ICE	T	x	—	x	Eng	Endo	H, C, V, L	x	PV, CSHP	x	—	—	x[a]	Sub-hourly	Commercial, open-source

Notes: S = steady-state, T = transient, Eng = engineering model, Sta = statistical model, Endo = endogenous, Exo = exogenous, H = heating, C = cooling, V = ventilation, A = appliances, L = lighting, E = electricity, DHW = domestic hot water, PV = photovoltaics, GSHP = ground-source heat pump, BIPV = building integrated photovoltaics, HP = heat pump, TES = thermal energy storage, CHP = combined heat and power.
[a]Coupled with generic optimization programs; [b]Coupled with Matlab & Polysun Inside (a Polysun plug-in).
Source: Sola *et al.*[151]

generation from PV, wind turbines and various types of energy storage (thermal and electrical storage with various control and management strategies). EnergyPlus runs simulations at a time step ranging from 1 to 60 minutes for a single day, an entire year or multiple years. EnergyPlus has comprehensive outputs, including summary reports of key inputs and monthly and annual performance of equipment and systems, as well as time-series reports of thousands of variables. EnergyPlus is a simulation engine — it takes several text files representing the building and systems (as well as weather data), runs the calculations and outputs several text files with various types of results.

EnergyPlus is open source and continues to add new features and improve its usability in the six-month release cycle. It has passed test cases defined in ASHRAE Standard 140.[144] Most of today's HVAC system types and controls can be modeled in EnergyPlus. Advanced users can use the Energy Management Systems feature to write scripts to overwrite the built-in calculation algorithms or implement a customized control strategy. EnergyPlus also has a co-simulation capability using the functional mockup interface. EnergyPlus can be used in co-simulation environments that couple the output from EnergyPlus with other simulations such as computational fluid dynamic (CFD) tools, occupant behavior tools (e.g., obFMU[65]), and daylighting rendering tools (e.g., Radiance). EnergyPlus runs on multiple platforms and has been adopted by major US vendors, including Autodesk (Green Building Studio), Trane (TRACE), Carrier (HAP) and Bentley (AECOM).

To support diverse user needs, middle wares have been developed that wrap around EnergyPlus. For example, the OpenStudio ecosystem, including OpenStudio SDK, OpenStudio App and the Parametric Analysis Tool, provides a programming interface and tools to ease vendors' adoption of EnergyPlus in their own software tools. The OpenStudio Building Component Library contains 300 OpenStudio measures (i.e., Ruby code scripts) that can be directly employed to evaluate performance of many energy conservation measures or building technologies. DOE also developed a suite of reference energy models covering most types of commercial and residential buildings across all climate zones,[143] which are used to support the development of ASHRAE Standards 90.1 and 189.1, as well as to evaluate building technology potentials at the national scale.

Built upon the EnergyPlus engine and the OpenStudio SDK, the Commercial Building Energy Saver (CBES)[66] is a web-based toolkit that provides a suite of features to guide building owners and engineers to evaluate retrofit measures, using a library of 100 building technologies, to reduce building energy use. Besides using EnergyPlus for detailed modeling, CBES also employs clustering and statistical data analytics to analyze smart meter data to identify low or no-cost operational improvement opportunities. Recent additions to CBES include features (e.g., PV, VRF+DOAS, electric battery) to support the design or retrofit of buildings to achieve zero net energy goals. CBES has a California version using California's Title 24 building energy efficiency standards and California's 16 climate zones data, and a US national version using ASHRAE Standard 90.1 and ASHRAE climate zones data. Both versions are free to use and available at CBES.lbl.gov and CBESPro.lbl.gov.

Built upon the CBES toolkit, CityBES[67,68] is a web-based data and computing platform that focuses on energy modeling and analysis of a city's building stock to support district or city-scale energy efficiency programs. CityBES uses an international open data standard, CityGML, to represent and exchange 3D city models. CityBES employs EnergyPlus to simulate building energy use and savings from energy efficient retrofits. CityBES provides 3D-GIS integrated visualization capability that can visualize a dozen metrics of building performance, including energy use, peak demand, water use, savings potential, retrofit payback, GHG reduction, energy cost savings and investment costs. CityBES provides a suite of features for urban planners, city energy managers, building owners, utilities, energy consultants and researchers.

3.6. Opportunities to improve energy efficiency and performance in homes

3.6.1. How to judge energy use — what metrics should we use?

Currently, most residential energy use metrics normalized by floor area. This is an approach that attempts to account for the energy use per service provided, and it seems reasonable that a bigger home provides more service. However, this approach leads to several problems. First, it rewards larger homes that actually use more energy because the heating and cooling loads scale with envelope area that increases less quickly than floor area. A small home that uses less energy requires more wall insulation, better windows and more efficient heating and cooling equipment than a large home to meet code requirements or comply with voluntary programs. Some local authorities, and recently the energy ratings provider, RESNET, are using various methods to address this, but it remains an important issue. There are other metrics that could be useful, such as an energy use per occupant (or number of bedrooms as a surrogate), and these need to be explored in energy codes, standards and programs.

3.6.2. *Electrification*

If we are serious about sustainability and reducing climate change effects, it is essential to use a metric of GHG reduction instead of simply focusing on reducing energy use. This implies that we need to use energy sources that are low or zero carbon dioxide content — primarily renewables. Since it is not feasible to capture carbon dioxide from buildings using gas or oil, buildings must operate only with electricity. In homes the major changes are for the gas end uses: heating, hot water, cooking and clothes drying, of which heating is the largest gas consumer. Heat pumps are now available that have good performance to low ambient conditions (–30°C) and demonstrated in field studies.[68] Similarly, heat pump water heaters (HPWH) are becoming an increasingly common way to heat water. Combined heating and domestic hot water heat pump systems are coming on the market and being used in high performance new and retrofit homes.[69,70,71]

For cooking, the most efficient option is to use induction cooktops. They also have the advantage of emitting less contaminants than gas (or electric resistance element) cooktops and are therefore better from an IAQ perspective. Electric clothes dryers are readily available and inexpensive, but efficiency gains can be made using heat pump dryers — with a trade-off for capacity and drying time. While some of these choices are currently niche options they are an essential part of reducing household carbon dioxide emissions, and will definitely become more important in the future.

Lastly on the topic of electrification are issues with peak demand. Homes contribute significantly to peak load for utilities, and developing ways to shift that load off peak or to reduce the peak demand are essential as we move to full electrification. Some HVAC loads can be shifted by several hours using the thermal mass of buildings, and research is underway to explore the use of active phase change technology for homes. Some loads, such as ventilation, can be shifted in time (and smart ventilation strategies are currently in development to do exactly this.[145] Other loads, such as lighting, cooking and plug loads offer much less of an opportunity to be time-shifted. The parallel electrification of personal transport will provide an opportunity to integrate the electricity storage in cars, motor-cycles and bicycles into a home's electrical system.

3.6.3. *New homes*

The new home construction industry has developed to the point where many mainstream builders are designing and constructing homes that use little energy and provide a healthy, comfortable indoor environment. There are many high performance home programs run by states, utilities and other entities that have been successful in achieving essentially zero energy homes for little or no cost increment. This has been achieved by being smarter about building homes, such as trading off the cost of more insulation or better windows against the lowered cost of heating and cool-ing systems that have to meet a lower building load, or using minimum framing that saves materials costs and labor while improving the energy performance of the wall assembly. California now requires that the mini-mum code compliant home you can build is a net zero energy home, i.e., it just needs a moderate solar PV system (4 kW for a typical home) to offset all its energy use. And the state has reached this point without requiring any great technological breakthroughs, just using simple, robust, well-proven approaches that the industry is mostly familiar with. This better use of existing technologies has been an important aspect of suc-cessfully achieving net zero energy homes because it lowers the risk inher-ent in innovation in an industry that is highly risk-averse.

To achieve net zero energy performance, all energy end uses must be addressed. Walls need to be well insulated beyond simply filling a wood frame wall cavity with insulation. This typically includes a layer of foam board or the use of spray-foam insulation that also minimizes thermal bridging and results in much more airtight construction. Windows are one area where recent and emerging performance upgrades, including vacuum fills and lightweight triple-pane systems, are aiming to roughly halve the energy lost through them. New homes are becoming much more airtight than existing homes. The 2018 International Energy Conservation Code[72] requires a maximum level of leakage of 3 air changes per hour at 50 pascals (ACH50) for most of the US. The LBNL Air Leakage Database (resdb.lbl. gov) contains more than 130,000 blower door tests of homes and shows that the US housing stock average is 15 ACH50. As shown in Fig. 27, this was reduced to about 7 ACH50 in homes built in the 1990s.[73] There has also been a matching reduction in the variability of home air leakage.

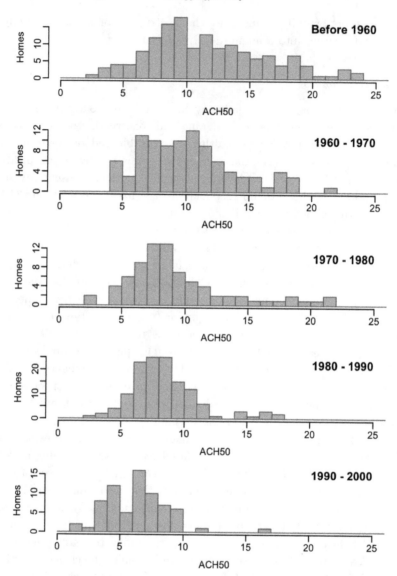

Fig. 27.   Changes in home air leakage.
Source: Chan *et al.*[73]

As the thermal performance of building envelopes has increased, the remaining loads of hot water, ventilation, lighting appliances and plug loads are becoming more important and represent a limit on low energy consumption that needs to be addressed. A key part of plug loads is the energy used even when the devices are not being used. The NRDC[74] estimates that this energy use costs the average homeowner \$165/year, or about \$19 billion nationwide. Suggested solutions include behavior (getting people to unplug devices), labeling products to indicate idle power consumption/energy use and improved minimum efficiency standards. Some of these are "builder installed loads" such as ventilation fans, carbon monoxide and smoke alarms, and GFCI controls. A survey of builder installed loads[51] found that an average new home uses 128 W continuously even before anyone moves in. One suggested solution in new construction is that a DC circuit could be provided in the home that would allow the elimination of the power supplies in these devices. Similarly, a key problem from the "internet of things" (IOT), where the electrical devices in the home are all communicating, is that all those smart devices (often relying on an internet connection) require energy to operate, and it is far from clear that the available savings can offset this additional energy. This will require some management in the future, possibly with IOT equipment interoperability and performance standards.

For hot water, heat pump water heaters (HPWHs) are currently the most efficient method, but their installation requires careful design, depending on climate: in cooling dominated locations they can cool the indoor air to heat water, but in homes where heating is bigger than the cooling load (the vast majority of homes) the extra cooling is not an advantage. In these homes HPWHs should be exchanging heat with the outside — requiring extra ducting and limiting the locations they can be placed in a home. Other issues with HPWH revolve around noise and the need to retrofit the home with new electrical wiring/service that substantially increases the cost. For the near future, high performance homes are likely to use on-demand gas water heaters in situations where HPWH are not the optimum solution (although, like HPWH, in retrofits their extra capacity may require gas line replacement with its associated cost increases). Another way to reduce hot water energy use significantly is to use the shortest piping runs, to reduce the wait time for hot water and the wasted hot water left in pipes at the end of each water draw. In new homes this can be achieved with changes to floor layout that group together the hot water end uses close to the hot water source. Another alternative is on-demand and/or point source water heaters that are likely to become more popular as we electrify all our end uses.

To go beyond net zero energy requires a greater investment in home design and construction. An example of this would be the Passive House approach, which requires much more insulation, more airtight envelopes (ACH50 < 0.6), and high performance windows to achieve its energy use targets. The original Passive House concept was to have a home that required no heating system at all, and instead using high performance envelopes and solar heating. Passive House principles began in Northern Europe and also have been developed for the US market (www.phius.org). They use solar heating combined with highly insulated tight construction that limits building space conditioning loads to less than 3,840 kWh/person/year of source energy. Although Passive Houses are rare and will continue to be so due to the complexity of their construction (but not necessarily cost — the premium is only about 5–10% in the US according to the Passive House Institute US [PHIUS]) — they are useful as a proving ground to show what can be done, and to show how extreme construction can be made to be risk free and acceptable to occupants. Passive House places an emphasis on indoor air quality as a selling point, with a great deal of attention paid to ventilating the home adequately. Passive homes in the US are designed to earn the US Environmental Protection Agency (EPA) Indoor airPLUS label (https://www.epa.gov/indoorairplus) for IAQ, as well as the DOE Zero Energy Ready certification. As a promotional tool, and to provide guidance for future passive homes, PHIUS maintains a database of certified homes, just like its European/International partner does. About 2,300 passive homes had been built in the US by 2018, and tens of thousands have been built worldwide.

Multifamily buildings have received less attention for reducing energy use. This is because there is often a disconnect between who pays for energy use (the occupant) and who would pay for an improved building (the building owner) — the "split incentive" problem. They also suffer from a lack of roof space when considering solar PV installations to get to net zero. However, several studies have measured energy savings from multifamily retrofits that have demonstrated their cost-effectiveness.[75,76] Many zero net energy multifamily buildings have been constructed, and some studies have shown remarkably low costs (a few hundred dollars per unit) for zero energy home approaches.[77]

3.6.4. *Existing homes*

Saving energy in existing homes is much more complex and expensive because it is necessary to interact with or remove an existing building component or system. There is also the inconvenience and disruption to the people already living in the home. Moreover, there are significant technical challenges. Moisture, for example, requires a comprehensive design approach that is easy in theory, but becomes difficult when implementing that approach on an existing building whose construction and building systems are currently in balance, and changing one of them could have serious consequences. This results in resistance in the energy retrofit community to relatively simple approaches such as insulating wall cavities due to fear of moisture issues. Because of the large variability in the performance of existing homes, several performance targets have been identified: 50% energy savings, equivalence to new construction, or an absolute energy use limit. All of these have been used effectively in the design of energy retrofits.[146] Even relatively expensive deep energy retrofits have been found to be cost-effective on average, but the range of costs are large and will need to be controlled if deep retrofits are to become more mainstream (Fig. 28). It is worth noting that these costs do not include externalities such as the social cost of carbon. If such externalities were included such retrofits would be even more cost effective.

Home improvement is a massive industry — typically about $250 billion a year in the US. Energy and IAQ upgrades do not even appear on the list of home upgrades according to analyses from the remodeling industry. The challenge is to get homeowners, banks, real estate

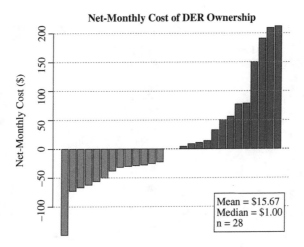

Fig. 28. Variability in cost effectiveness of deep energy retrofits.
Source: Less and Walker.[146]

professionals, insurance companies and home appraisers to recognize both the monetary value and the value placed by homeowners on good IAQ and energy savings. This concept of rating an asset (the home) for IAQ faces challenges regarding what metrics to use for comparing different homes that will work for a range of potential users. Metrics are currently under development that include IAQ-related health, odors and moisture that could allow homes to be scored in the same way that homes are scored for energy use (e.g., the Home Energy Rating System [HERS] Index at www.resnet.us) that is used to rate energy use in about 40% of all new homes in the US (Fig. 29).

One viable approach when retrofitting a home is to not do everything at once, but to stage improvements over several years. This reduces the financial burden on homeowners and allows incremental steps that are individually less disruptive than doing everything at once. An associated concept is loading order, where the order in which retrofits are carried out is key in ensuring success. One example is to air seal a home before insulating it because once the insulation is in place it is difficult to access the air leaks to seal them. Another is to reduce building loads before installing on-site generation.

Some new approaches to home retrofits are being developed. One is to completely re-clad buildings (possibly with integrated HVAC systems) using panelized construction techniques where the new building envelope is built in a factory and assembled from a few pieces on site. An example

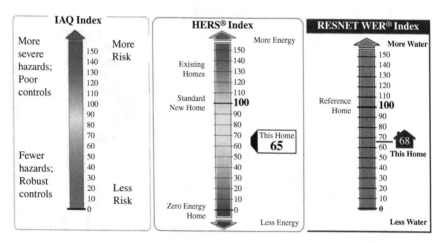

Fig. 29.   Illustration of the new IAQ Index that allows for ratings similar to those for energy and water use.

of this is the "energiesprong" system being used in Europe. Another approach is to not remove existing HVAC systems, but instead install a heat pump that covers 95% of the heating and cooling load, and retaining the old system as backup for the final 5% of the load.

3.6.5. *Ventilation: Indoor air quality and energy*

Ventilation is essential for dilution and removal of indoor air contaminants from people, furnishings, construction materials, etc. Ventilation air needs to be conditioned, and in a high performance home whose other end uses have been minimized, the energy to condition ventilation air becomes increasingly important. Recent innovations in "smart ventilation" have focused on ways to maintain IAQ while saving energy. A key to smart ventilation is to use controls that can time shift to when an energy penalty is lower, take credit for operation of other fans (kitchen, bath and dryer exhaust, and economizers in some climates), and/or ventilate less when occupants are not home but not change exposure to contaminants.[78] In Europe, demand controlled ventilation (usually based on CO_2 and H_2O as surrogates for occupancy/odor) is a core part of energy efficient building standards[79]).

Substantial evidence suggests that with careful design and operation, high performance homes may improve occupant health and reduce pollutant levels.[80-86] The sealing of crawlspaces to improve energy performance also leads to lower moisture levels, mold and spore transmission to the inside of homes.[87] Air sealing reduces pollutant entry from attached garages.[88] Other energy-related home improvements that contribute to these results are: air sealed HVAC ducts to limit transport from attics, crawlspaces and garages; mechanical ventilation to provide more consistent outdoor air exchange, and combustion appliance testing, sealed combustion appliances, or all-electric appliances to remove sources of combustion-related contaminants.

The key mantra for IAQ in high performance homes is: Build Tight — Ventilate Right. But this raises some questions: How tight is tight enough? And how much ventilation is needed?

For new homes the airtightness level used in the International Energy Conservation Code$^\Delta$ (IECC) (3 ACH50) captures about 80% of savings and is a reasonable target. For higher performance homes, a lower target of 2 ACH50 can be used to capture about 90% of potential savings in non-mild climates. The 0.6 ACH50 used for passive homes is not necessary for mass market, market rate housing. In retrofitting homes the target should be at least a 50% reduction in air leakage, or less than 5 ACH50.

Ventilation rates for minimum acceptable IAQ in homes in the US are set by ASHRAE Standard 62.2, and these result in air change rates of about 0.3 to 0.35 air changes per hour. The ASHRAE standard specifies ventilation rates required to dilute indoor contaminants as well as exhaust system specifications for kitchens and bathrooms. It is possible to do better than this minimum, either for dilution of outdoor contaminants, removal of contaminants by filtration, or selecting ventilation systems that minimize energy use. For example, in many US climates a heat recovery ventilator or energy recovery ventilator can save 70% or more of the heating and cooling energy associated with ventilation. Recent work using disability adjusted life years (DALYs)[89] has identified the key contaminants of concern for health: small particles ($PM_{2.5}$, particles less than 2.5 microns in diameter), formaldehyde (primarily from building products and furnishings), contaminants that come from combustion (cooking, candles, incense, secondhand smoke) and building materials, furnishings, and household cleaning products (Fig. 30). The large range of exposures is a combination of the range of emission rates and health impacts across the various studies used in developing the analysis. The biggest ranges are in the health impacts — both within studies and between studies.

Filtration can be effective at removing particles, and the ASHRAE Standard 62.2 has recently been amended to include a credit for filtration. Formaldehyde can also be reduced by selecting low formaldehyde building products and furnishings. Figure 31 shows how increasing ventilation rates and selecting low-emitting materials lowers formaldehyde concentrations measured in homes. Combustion products are difficult to remove, and source removal (e.g., using a good kitchen range hood) and dilution through ventilation is the only practical approach. Increasing attention is being paid to outdoor air contaminants that can enter a home. More recently, work on outdoor particles has shown that a tight building envelope can be good filter, with a typical new home being equivalent to a MERV 13 filter on a supply system.[90] Although current IAQ standards do not require filtration of outdoor air for supply or balanced systems, we are likely to see requirements in the near future.

Radon has been well known for many years as a contaminant with serious health impacts. Radon control systems are often required in parts of the country with high radon levels. The EPA produces radon maps that show where levels of radon are high.

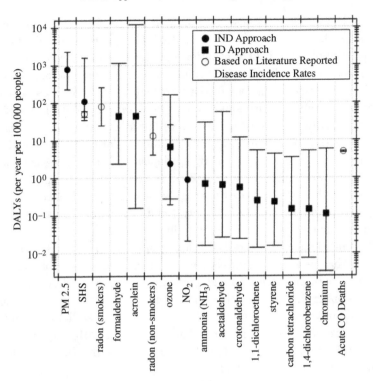

Fig. 30.   Contaminants of concern in homes.

Source: Logue *et al.*[89]

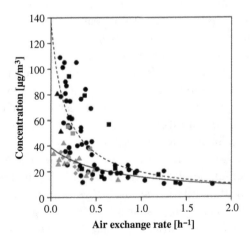

Fig. 31.   Changes in indoor concentrations of formaldehyde with air change rate and selection of low-emitting materials (red symbols).[91]

There are low-cost sensors coming to the market that will make it possible in the near future to get away from the generic contaminants used as the basis of most IAQ ventilation standards, and instead ventilate to control contaminant levels. These sensors also give us the potential to have ventilation systems that can respond to high levels of outdoor contaminants and, combined with smart ventilation controls, control indoor contaminants without introducing excessive outdoor contaminants. This has great potential, but there are concerns about ventilation system designs that are not robust, i.e., being highly dependent on sensors that do not drift or fail, or require more maintenance, compared to a simple exhaust fan. Current low-cost sensors have highly variable performance, with some being acceptable for ventilation control and others not.[92] In the future we will need standardized testing and rating for these devices.

Commissioning of homes is becoming more important, either as a way to show that performance levels have been met for code minimums or to demonstrate compliance with high performance home standards. Generally, commissioning of homes is done via a set of diagnostics with a focus on air flows and visual observation (e.g., insulation installation practices). Diagnostics include envelope air leakage (the IECC 2018 requires a maximum level or air leakage of 3 ACH50 for most of the US), duct leakage (typical upper limits are 6% of the total system air flow), and ventilation air flows for IAQ. Overall, trends are toward much tighter envelope construction in new homes, and air tightening is usually one of the first targets for home energy retrofits; it has other benefits, such as reducing noise transmission, transport of outdoor air contaminants (such as particles) into the home, and drafts, leading to increased comfort. Of these diagnostics, the measurement of envelope leakage is by far the most prevalent, followed by duct system leakage. The measurement of ventilation air flows is required by the ASHRAE 62.2 national ventilation standard, but in practice it is rarely done. Note that some ventilation systems (e.g., exhaust fans) are much easier to measure than others (such as supplies connected to central forced air systems). We simply do not have the appropriate technology for the latter.

In a tight high-performance home (say, below 1.5 ACH50, which is one-half the maximum level set in the IECC) the occupants are dependent on their mechanical ventilation system. It is important that it be maintained. This includes checking its air inlets and changing its filters. Recent studies of mechanical ventilation in single-family homes has shown that most occupants know nothing about their ventilation system, cannot tell if it is operating and do not know whether they should operate it.[93-94] Combined with poor labeling, this leads to most homes having their ventilation systems turned off. This was not so critical in older leaky homes that had plenty of natural infiltration, but it is a growing concern for more airtight energy efficient homes. Considerable work is needed to improve public awareness and labeling of ventilation systems, particularly in high performance homes.

During the last few years many areas of western US and other regions have experienced extreme smoke and poor outdoor air quality from large fires. This problem introduces a growing problem of how to ventilate homes during such periods. While the general policy is to close windows and run air conditioners if available, we need more research and better information for homeowners in order to inform them about which type of filters to purchase and how to understand their ventilation systems.

A final comment about residential buildings in the US is that the US government has made significant investments in energy efficiency standards over the last few decades that will continue to provide energy savings for years to come. The Building Technologies Office (BTO) implements minimum energy conservation standards for more than 60 categories of appliances and equipment. As a result of these standards, American consumers saved $63 billion on their utility bills in 2015 alone. By 2030, cumulative operating cost savings from all standards in effect since 1987 will reach nearly $2 trillion. Products covered by standards represent about 90% of home energy use and thus we mention it in this section. However products covered in standards also account for 60% of commercial building use and 30% of industrial energy use.[96]

3.6. *Opportunities to improve energy efficiency and performance in commercial buildings*

3.6.1. *Building systems: control and analytics*

Historically, energy efficiency has emphasized the performance of building equipment, that is, systems and devices that provide a given service level with lower levels of energy inputs. For example, the transition from incandescent to compact fluorescent and now LED lighting sources has improved efficacy from approximately 15 lumens/W to nearly 100 lumens/W.[97] Similarly, the efficiency of packaged unitary air conditioners has nearly doubled since the early 1970s.[98] While the pace of improvement varies, analogous advances have been observed for appliances, windows, building envelopes and electronics. These asset-level gains in efficiency are commonly secured at the time of new construction, through natural replacement, or through capital retrofit investments.

Beyond these equipment-focused gains, there is tremendous opportunity to apply advanced controls and continuous analytics to obtain even deeper performance improvements. Fernandez *et al.* (2017) estimate annual national savings potential of 29%, just from the proper use of advanced controls, sensing and diagnostics in the US commercial buildings sector.[99] These low-/no-cost solutions complement equipment-level efficiency to help realize maximum building energy performance, as well as associated benefits of reduced utility expenditures, improved comfort and indoor environmental quality, and grid integration. They are further discussed in the following, with illustrations from the commercial buildings sector.

Today's best practice in building controls can be characterized as providing "static" energy efficiency. Sequences of operation are designed to implement scheduling, off-hours conditioning setbacks and reset of control parameters that influence system energy consumption. Coordination across end uses such as lighting and HVAC may be implemented in an open-loop manner. In this static paradigm, the same control strategies will be used from one day to the next, with some adjustment for seasonal changes.

An emerging, leading-edge approach is to use model-predictive control strategies to *dynamically* modify control actions based on forecasts of required service levels (system load), weather, occupancy and other factors. Model-predictive control can be further enhanced by allowing the form of the underlying model to evolve as conditions change — this is referred to as *adaptive control.*

In moving from standardized sequences to model-predictive and adaptive control it becomes possible to operate buildings in a way that optimizes multiple objectives and exploits the interactive effects among end use systems. Goal-based policies can then be implemented to, for example, control HVAC in concert with lighting and plug loads while maximizing occupant preferences and minimizing total utility costs or energy consumption.

Tightly coupled with controls are monitoring and analytics technologies that provide continuous performance feedback and diagnostics to inform operators and service providers when failures or suboptimal operations occur. Energy management and information systems comprise a broad family of tools and services to manage and sometimes control commercial building energy use. These technologies offer a mix of capabilities to store, display, and analyze energy use and system data, and in some cases, provide control. They include energy information systems that focus on interval meter data analysis, central building automation systems (BAS) for traditional HVAC control, and fault detection and diagnostic (FDD) tools. Most recently, the energy management and information system (EMIS) market has also begun to deliver automated system optimization (ASO) technologies that provide model-predictive supervisory optimization, bringing model-predictive control out of research and development and into commercialized products.

While the various flavors of EMIS (Fig. 32) are converging in some cases, it is generally the case that fault diagnostics technologies rely primarily upon data from building automation systems (e.g., HVAC system temperatures and component operational status), and energy information systems rely primarily on data from whole-building and submeter level electricity and gas meters. HVAC optimization products typically use both BAS and meter data.

Independent of the nuance in technology types and the data that they leverage, EMIS users are making cost-effective use of the technology to improve building efficiency on the order of 5–10% through a focus on

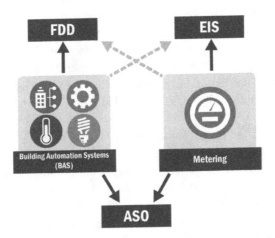

Fig. 32. Typical data sources for various EMIS technologies, including fault detection and diagnostics, energy information systems, building automation systems and automated system optimization.

operational as opposed to capital improvements.[100] Moreover, the technology is continuously advancing, particularly in the areas of multi-end use and model-predictive control.

Software-based analytics technologies (and to a lesser degree, optimization technologies) are inherently process tools that must be integrated with business and operational practices. Awareness and buy-in by operational and energy management staff is critical to realizing maximum value. Consider for example, the need to ensure that smart optimized controls are not overridden and that corrective action is taken promptly once problems are identified through the building's continuous analytics.

While this chapter has drawn heavily upon examples in the commercial building sector, analogous opportunities for operational efficiency, as opposed to equipment-level or asset-based efficiency, exist in the residential and industrial sectors. Advanced controls and analytics have a unique role to play in advancing energy efficiency for process-driven facilities, residences, industrial energy managers and homeowners.

3.6.2. *Delivery systems: energy management, operations, and maintenance*

The previous section presented non-capital operational opportunities for energy efficiency. Here, we discuss how critical considerations of these process technologies can be integrated with energy management and operations and maintenance practices for maximum impact.

Analogous to the case of building systems, the energy management practice has tended to focus on capital projects and the installation of more-efficient equipment assets. Similarly, operations and maintenance (O&M) practice has historically been driven by a focus on malfunctioning equipment, trouble calls, and scheduled or failure-driven equipment servicing.

In contrast, building commissioning is a proactive and systematic process for ensuring that buildings meet their designed energy and operational performance. There are several types of commissioning as shown in Fig. 33. It can be conducted for either newly constructed or existing buildings; it can incorporate information technology (monitoring-based commissioning); and it can be performed on an intermittent or continuous basis (ongoing commissioning).

As the benefits of controls and analytics technology are increasingly recognized, they have become some of the most frequent and effective

Fig. 33. Commissioning can vary along three dimensions: (1) life cycle (new vs. existing buildings), (2) type (conventional vs. monitoring-based), and (3) frequency (intermittent vs. continuous).

measures undertaken in commissioning. Monitoring-based commissioning that includes fault detection and other diagnostic technologies provide a link between advanced O&M and energy management.

A compelling evolution in the industry is seen in the expansion of market delivery of FDD through third-party service providers using the tools as a way to provide added value to their customers. As illustrated in Fig. 34, these services may cover a spectrum of activities. This is in contrast to earlier models that relied on an organization's technology use by in-house staff or via analysis services procured from FDD software vendors. In-house direct use is the concept that facility staff are the primary users of operations analysis tools. Analysis-as-a-service is the concept where a third-party company has a contract to provide information and analytics about operational issues and facility performance.

Business process frameworks for energy management, such as the ISO 50001 Standard (ISO, 2018) and *Energy Star Guidelines for Energy*

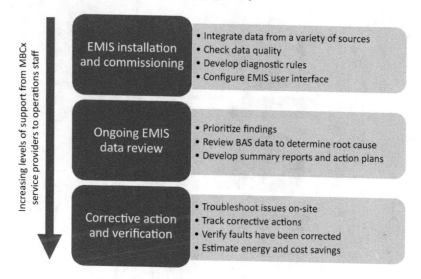

Fig. 34. A spectrum of analytics-focused activities that service providers may offer their customers.

Management[101] foster continuous improvement through a *plan-do-check-act* cycle. They provide a unifying framework through which in-house and third-party services can be integrated with advanced technology to deliver operational energy efficiency.

3.7. *Benchmarking and energy transparency*

"How does my building compare to its peers?" Building energy benchmarking is a means to answer this question for building owners, operators and service providers. It is often the first step in screening for efficiency potential. It also has become an increasingly popular policy mechanism for motivating energy efficiency actions, and there are now several voluntary and mandatory programs across the US and other countries. This section will first cover the technical aspects of benchmarking and then review the market and policy applications of benchmarking.

3.7.1. *Benchmarking methods and tools*

From a technical perspective, benchmarking is straightforward in principle: a building is compared to its peers using one or more quantifiable metrics.

For example, an office building could be compared to other similar office buildings using a metric such as total annual site energy use per unit floor area, or site energy use intensity (EUI). The technical complexity comes from how to select a peer group for comparison so as to create an "apples to apples" comparison of efficiency. This is accomplished by normalizing the data for the unique aspects of individual buildings that may cause their energy use to be higher or lower for reasons other than energy efficiency, such as different business operation hours.

There are a range of different approaches to normalization, and these generally fall into three broad categories[102]:

- *Simple data filtering*: Select a peer group of buildings that meet certain filter criteria, such as geographical region, floor area range, etc. The advantage of this approach is that it is transparent and straightforward. However, it requires a dataset large enough to allow it for "slicing and dicing" to meet a range of filter criteria.
- *Multivariate regression*: In this approach, a multivariate regression analysis yields an equation that relates the normalizing parameters to the metric of interest. This equation is then used to normalize the value of the metric for each building. This approach works well, provided there is a large enough representative dataset (including normalizing parameters) to run a regression.
- *Simulation*: In this approach, a simulation model is used to calculate a benchmark (typically representing an "ideal" case) against which the actual energy use can be compared. The model accounts for the relevant normalizing parameters.

The ENERGY STAR Portfolio Manager tool[103,104] is far and away the most widely used benchmarking tool in the US. It can also be used for Canadian buildings. The tool can be used in areas that have climates similar to the US. It provides a score from 1 to 100, and buildings scoring 75 and higher are eligible to receive an ENERGY STAR label. There is a separate scoring model for each building type, and most building types use the US Commercial Building Energy Consumption Survey[105] as the peer group for comparison. The scoring models use a regression-based approach (Fig. 35) to normalize for significant variables (e.g., for office buildings, it normalizes for floor area, weather, operation hours, number of people and number of computers).

Energy Efficiency

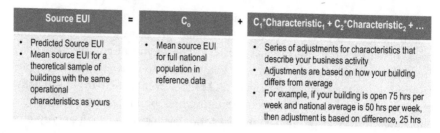

Fig. 35. Form of regression equation used to normalize for building characteristics in ENERGY STAR.

Source: ENERGY STAR.[103]

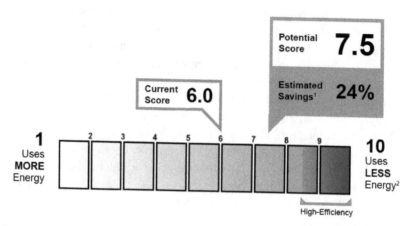

Fig. 36. A DOE Asset Score provides a rating for the fixed assets in a building and efficiency improvement potential.

Other tools support different types of benchmarking. ASHRAE's BuildingEQ tool[106] can be used to benchmark the actual energy use of a building as well as the "as designed" energy use of a building using simulation-based method. The DOE Asset Score[107] tool uses a simulation-based methodology to benchmark the physical and structural energy efficiency of commercial and multifamily residential buildings on a 1–10 scale (Fig. 36). The tool provides a current score and a potential score. The potential score considers practical retrofits that might be cost effective in a lifecycle cost analysis. That is why the potential score might be lower than the high efficiency case.

The DOE Building Performance Database[108] is a crowdsourced database that includes measured energy performance data on more than a

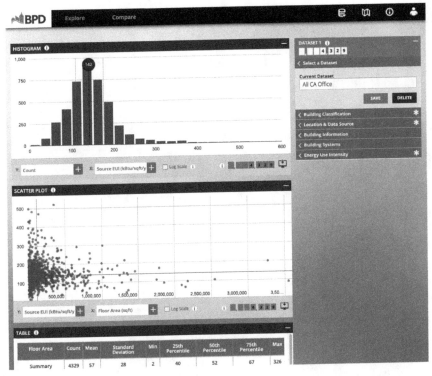

Fig. 37. The DOE Building Performance Database allows users to compare their building to a customized peer group using a range of various metrics and filters.

million buildings from more than 70 private and public-sector datasets. It enables users to compare a building to a customized self-selected peer group (analogous to real-estate "comps") using a range of various metrics and data filters (Fig. 37).

The most common metrics used for benchmarking of nonresidential buildings are site EUI, source EUI and ENERGY STAR score. Some tools also offer metrics for system-level benchmarking. For example, the Laboratory Benchmarking Tool[109] enables users to benchmark using metrics such as ventilation system W/cubic feet per minute, lighting W/ft^2, among others.

3.7.2. *Market and policy applications*

Market uptake of benchmarking has grown significantly over the past two decades; 40% of the US nonresidential building stock has been

U.S. City, County, and State Policies for Existing Buildings: Benchmarking, Transparency, and Beyond

Fig. 38. US state and local benchmarking policies.
Source: IMT.[152]

benchmarked in the ENERGY STAR Portfolio Manager. Benchmarking is considered to be a key first step in conducting an energy audit and is a mandatory requirement in ASHRAE Standard 211 for Commercial Building Energy Audits.[110]

More notably, over the last few years there are a growing number of state and local ordinances requiring benchmarking (Fig. 38). Approximately 11 billion ft^2 of floor space in major real estate markets is now subject to benchmarking requirements (Fig. 39). Washington recently passed legislation that will require existing buildings to meet certain performance requirements based on energy benchmarking.

The European Union's Energy Performance of Buildings Directive (EPBD) requires energy performance certification of buildings across EU member states, although each member state can implement it in their own way. Some schemes included both an asset and operational rating for buildings.

The rationale for these benchmarking and transparency requirements is that such information can then be used by market actors such as owners,

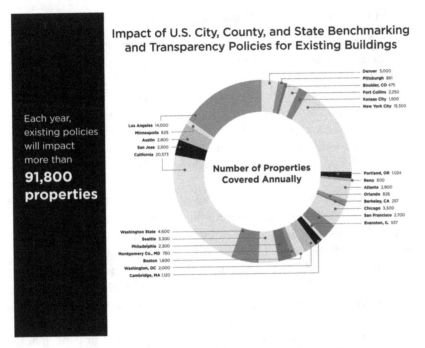

Impact of U.S. City, County, and State Benchmarking and Transparency Policies for Existing Buildings

Each year, existing policies will impact more than

91,800 properties

Number of Properties Covered Annually

Denver 3,000
Pittsburgh 861
Boulder, CO 475
Fort Collins 2,250
Kansas City 1,500
New York City 15,300

Los Angeles 14,000
Minneapolis 625
Austin 2,800
San Jose 2,500
California 20,573

Portland, OR 1,024
Reno 500
Atlanta 2,900
Orlando 826
Berkeley, CA 257
Chicago 3,500
San Francisco 2,700
Evanston, IL 557

Washington State 4,600
Seattle 3,300
Philadelphia 2,300
Montgomery Co., MD 750
Boston 1,600
Washington, DC 2,000
Cambridge, MA 1,120

Fig. 39.   Floor area covered by benchmarking ordinances in the US.
Source: IMT.[153]

tenants, investors, etc. The expectation is that poor performers will be motivated to improve their performance and that efficient buildings will be treated preferentially by the market. Over the past few years, a number of studies have looked at the relationship between energy efficiency and the financial value of buildings.[111] Zhu *et al.* documented a review of 39 peer-reviewed journal papers on this subject. The report showed considerable evidence that green building certifications correlate with increased economic value, primarily via higher rents and sales prices. Estimated rental premia were generally about 5% for LEED and ENERGY STAR certifications, but in some cases ranged up to 20%. Many studies revealed increased occupancy rates in certified buildings, though a few showed mixed results in this regard. Several papers found that green buildings were fetching higher prices than buildings with identical net operating incomes. These findings suggest that energy transparency can in fact affect the market, although there is a still a significant need to build the evidence and incorporate energy efficiency into building valuation as routine practice.

4.   China and India: Two Emerging Giants

China and India are two of the most populous countries in the world.[2] Each is home to nearly a fifth of the world's population. However, they have experienced different patterns of economic development, and as a result, China consumes four times more modern energy[d] in buildings than India does. India is still relying heavily on biomass as a source of energy consumption for cooking, notably in rural areas where households consume twice the level of energy consumption of urban households due to the very inefficient use of wood for cooking. This also poses significant health issues.[112,113]

Still, buildings in China currently use less energy than many of the Organisation for Economic Cooperation and Development (OECD) countries, but they are not as efficient as buildings in these countries. Unlike commercial buildings in the US, heating and cooling in Chinese commercial buildings is provided for only part of the building space, and may not be provided all day. Chinese buildings are built to provide a much lower thermal comfort level than those in the US, resulting in indoor temperatures often below 13°C in the winter.

Fast urbanization and population growth has resulted in poor quality construction and envelope materials in China. Residential urban buildings are mostly multifamily and high rise buildings, which is quite different from the US residential building stock. In northern China, space heating is provided by large, centralized district-heating systems, while southern China traditionally has no heating. Cooking and hot water consume a great deal of fuel. In rural areas that fuel is biomass (which is also used for space heating); whereas in urban areas, it could be coal and natural gas. All these energy uses will transition to commercial fuels as urbanization continues.

China's building sector has grown at an astounding pace during the past two decades and it now accounts for about 20% of the national primary energy consumption.[113] More than 2 billion m² of new residential and commercial building floor space were added annually over the last decade.[114–116] Construction of urban residential buildings has accounted for most of this growth, driven simultaneously by income growth and China's burgeoning urban population demanding more living space. More than 350 million new residents have been added to Chinese cities

---

[d]By *modern energy*, we mean all energy carriers except biomass that is mostly used for cooking.

since 2000. At the same time, urban residential floor space has nearly quadrupled, from 9.6 m^2/person in 2000 to 36.6 m^2/person in 2016, while urban household size has declined slightly, from 3.1 persons/household in 2000 to 2.9 persons/household in 2012.[117]

Over the next three decades, Chinese building floor space is expected to continue to increase, with growth driven by new urban residential and commercial construction. Studies at LBNL suggest that, by 2050, urban residential floor space could reach 46 m^2/person in China, and the total floor space could reach 50 billion m^2, as shown in Fig. 40. At the same time, the total rural residential floor space would decrease, from 22 billion m^2 in 2010 to only 14 billion m^2, as the share of rural population shrinks from the current 41.5% to only 22% by 2050. During the same time period, the total commercial floor space is expected to double, from 12 billion m^2 in 2010 to 23 billion m^2 in 2050, driven by expanding tertiary economic sector and a decline in industrial activity.[118]

Although there are no national energy consumption surveys for buildings in China, surveys conducted by universities and research institutes have found that heating (particularly district heating in the colder northern climate zones), cooking and water heating are the three largest end uses in residential households. They account for 54%, 23% and 14% of the

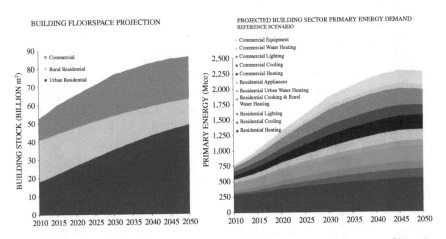

Fig. 40. Projected floor space projections and end-uses for energy use in China in buildings.

Note: Mtce is million metric tons of coal equivalent, the standard unit for energy in China. 1 Mtce = 29.27 million gigajoules.

Source: This data was based on work conducted in ERI, LBNL, and RMI[116] and Zhou *et al.*[114]

total energy consumption, respectively.[119] Electricity use for cooling and other household appliances account for only 9% of residential energy consumption, but both end uses are expected to rise with growing appliance ownership, especially for electronics and miscellaneous plug loads, as well as a greater demand for cooling, supported by rising incomes. By fuel type, the largest shares of residential energy consumption were met by district heating (primarily supplied by coal, with a smaller portion from natural gas), natural gas, liquefied petroleum gas and electricity, followed by direct use of solar, coal and firewood.[120]

Increasing demand for future energy services in China will come from both residential households with increasing income levels and China's expanding commercial sector, although current average energy intensity in China's residential and commercial buildings is still low compared to international levels. For residential buildings, the urban cooking and water heating intensity was only one-quarter of the average Japanese level in 2000. The average final energy intensity in commercial buildings is only one-third of US and Japanese levels.[122] As a result of the lower levels of thermal integrity in building envelopes in China compared to developed countries, as well as lower thermal comfort conditions compared to other developed countries in Asia, building energy consumption is expected to increase significantly in the future.[123]

India's forecasted economic growth suggests that a growing middle class is emerging with aspirations for a better quality of life.[124,125] As household incomes rise, consumers will purchase more energy-using assets like appliances. The penetration of air conditioners in particular is likely to increase rapidly, as India is one of the hottest countries in the world, with cities like Chennai, Mumbai, Kolkata and Delhi ranked among the top 10 hottest large cities globally, as measured by cooling degree days.

In its Demand Resources Energy Analysis Model (DREAM) models for China and India, LBNL provides descriptive quantitative projections of energy demand based on exogenously determined drivers and technologies penetration.[126,127] Figure 41 shows the historical saturation of three major types of household equipment in urban Chinese and Indian households, and that assumptions of India ownerships in 2050 will match China's 2015 levels. Projections from LBNL's DREAM models for China and India forecast building energy consumption to nearly triple in both countries over the next 30 years, driven by increasing demand for heating in China and cooling in India (Fig. 42).

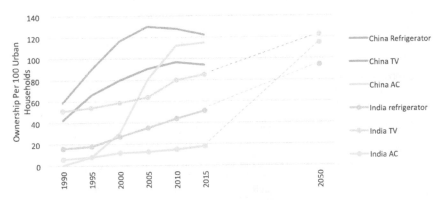

Fig. 41. Historical and projections of urban household equipment penetration in China and India, 1990–2050 (de la Rue du Can *et al.*, 2019).

4.1. *Energy efficiency policy*

To address the expected rise in building energy consumption, China has adopted a comprehensive set of building efficiency policies that include mandatory building energy efficiency codes, mandatory and voluntary building energy labels, and national and regional incentives for high-efficiency equipment and low-energy building designs.[127] Chinese building codes have been implemented in phases for different climate zones for residential buildings since the late 1980s, as have national standards for public buildings in 2005 and 2015. These standards primarily focus on reducing heating and air conditioning energy intensity by a targeted level, relative to inefficient baseline buildings from the 1980s. In addition, China's Ministry of Finance, along with some key municipal governments, have also provided financial subsidies for various building energy conservation measures (including retrofits), as well as renewable energy building applications including solar photovoltaics, ground source heat pumps and integrated solar thermal technology.[128] For green buildings, in particular, China has developed a national design standard, a green building evaluation standard and labeling system, and urban targets for the share of new green buildings construction.[129]

As indicated by the direction of recent building policies, future opportunities for energy savings in Chinese buildings include: lightweight and more durable construction materials, improved construction methods such as prefab, passive and integrated building design to reduce heating and cooling needs, adoption of highly efficient and environmentally friendly appliances and equipment (especially cooling), and use of renewable technologies in buildings.

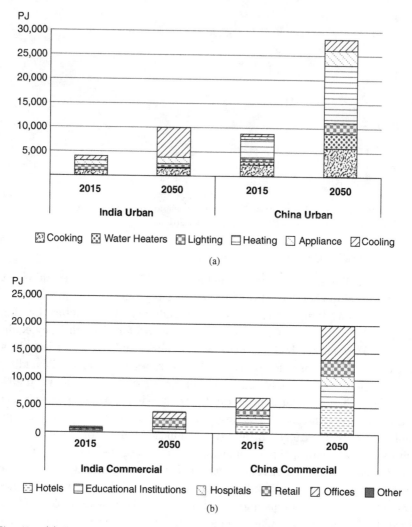

Fig. 42. (a) Residential and (b) commercial energy consumption forecasts for India and China.

Source: LBNL DREAM model.

4.2. China's new phenomenon of fuel switching to improve air quality

The use of lower-quality, highly polluting dispersed or scattered coal, particularly for distributed heating applications in northern China, has become a major policy concern in China. In the largest metropolitan areas of Beijing-Tianjin-Hebei (Jing-Jin-Ji), dispersed coal use, primarily for rural heating during the heating season, is estimated to be responsible for 50% of air pollution.[129] National and regional policies have been adopted over the last few years to curtail the use of dispersed coal, including setting targets for alternative heating technologies during the heating season. This specifically includes shifting from coal to natural gas and electricity for heating, leading to residential heating technology choices. Air quality concerns in urban areas also have led to rapid growth in the use of air filtration systems, as well as increased need for ventilation.

Given that most of the appliance stock is yet to be purchased in India, significant opportunities exist to minimize electricity consumption by increasing energy efficiency. Minimum energy performance standards in India are already mandatory for the following residential equipment: fixed-speed room ACs, refrigerators (frost free and direct cool), fluorescent tube lights, distribution transformers, electric water heaters and TVs. Standards for variable speed ACs and ceiling fans are under consideration.[131,132] India's government launched the first Energy Conservation Building Code (ECBC) for new commercial buildings[e] in 2007 and published a revision in 2017. ECBC-compliant buildings are 20% more efficient than conventional buildings, with additional savings for ECBC+ (30–35%) and for Super ECBC buildings (40–45%).[133] However, implementation varies across states and depends on resource capacity for verifying compliance. These programs could be expanded and reinforced to increase efficient use of energy in India.

4.3. *Growth in air conditioning load, affordability and urbanization*

As emerging economies located in hot climates become wealthier, the demand for air conditioning is expected to grow rapidly. Climate change is also causing warmer temperature and motivating greater adoption of air conditioning. The IEA projects global energy demand from ACs to *triple* by 2050, requiring new generation capacity equivalent to the current combined electricity capacity of the US, the EU and Japan.[14] The global stock of air conditioners in buildings is projected to grow from 1.6 billion today to 5.6 billion by 2050. China, India and Indonesia will together account

---

[e]Buildings having a connected load of 100 kW or contract demand of 120 kilovolt-ampere (kVA) and above.

for half of the total number of new air conditioned buildings. The additional electricity demand needed to power this new equipment will put tremendous pressure on the power sector of emerging countries by contributing significantly to increasing peak electricity demand.

Improving energy efficiency of ACs in parallel to an HFC transition could double the climate benefits of the Kigali Amendment to the Montreal Protocol. Shah[134] and colleagues[f] quantified the energy and climate benefits of leapfrogging to high efficiency in tandem with the transition to low-GWP refrigerants for room air conditioners and found that shifting the total projected 2030 stock of room ACs to more-efficient ACs with lower GWP refrigerants would globally save between 340–790 gigawatts (GW) of peak load and avoid up to 25 billion tons of carbon dioxide in 2030. Figure 43 shows the combined climate benefits of refrigerants and energy efficiency transitions in China, India and Indonesia. The study assumed an efficiency improvement of 30% over an average baseline of 2.9 EER (energy efficiency ratio). This is a conservative estimate, as an efficiency improvement of more than 50% is possible based on commercially available models, some of which are sold at competitive prices.[134]

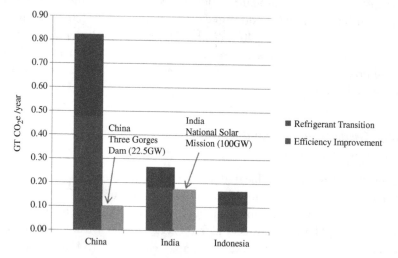

Fig. 43. Estimated annual emissions abatement potential of AC stock in 2030 over AC lifetime.

Source: Shah *et al*.[134]

---

[f]Buildings having a connected load of 100 kW or contract demand of 120 kVA and above.

New, stronger and additional policy interventions are needed to lock down these potential energy savings and related emissions reductions. Standards and labeling can help transform the market by raising the "floor" of efficiency, complemented by clear labels to best inform consumer purchase decisions and incentive programs to raise the "ceiling" of AC efficiency by accelerating the penetration of highly efficient technologies.[136]

5. Critical Needs and the Changing Energy Landscape

5.1. *Integration of building loads and the electric grid*

5.1.1. *Demand response in the US*

Since buildings represent nearly three quarters of all US demand for electricity, major systemwide efficiencies can be achieved by integrating smart building energy management with modern electric supply system dispatch (smart grids). The benefits of doing this are so significant that the electric utility industry has shifted from partnering with energy consumers solely to reduce energy use to building partnerships that jointly manage supply and demand to optimize overall system performance. These new partnerships move beyond static energy efficiency to dynamic energy management.

During the past few decades, investment has grown in technologies and programs to reduce electricity demand on days when the electric grid is stressed and near capacity. These are typically hot summer days or cold winter mornings. Peak power capacity constraints can result from rapid growth in a region's electric demand, limits on the ability to build new generation, or unplanned power plant disruptions. Traditionally, electric utilities have been forced to react to these problems by purchasing more electricity — which is usually very costly — to avoid blackouts or brownouts.

Many utilities are exploring how *demand response* (DR) can be a low-cost way to address these problems. These strategies provide financial incentives for customers who reduce their electricity use during DR events. Demand response also can be used to improve the integration of variable renewable electricity supply systems. Figure 44 summarizes DR objectives, communication data models, automation approaches and control strategies.

Much of the historical DR has been achieved through manual means, with utilities calling or faxing large customers to drop load during emergency events. Today, however, utilities are moving quickly to automate

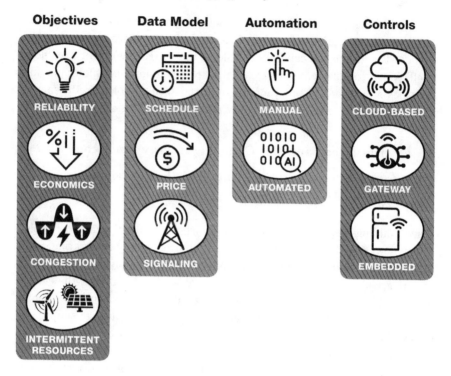

Fig. 44. Objectives for and methods of demand response.

the process. Table 6 lists the DR capacity that exists in the US. Many utilities have installed switches on air conditioners that enable US utilities to drop 3.4 GW of load during hot summer days. Customers receive financial incentives to participate. These programs can be lucrative to utilities, but challenging for their customers if the AC is turned off on the hottest days of the year. Some utilities turn off AC units for 4 to 6 hours in exchange for a one-time financial payment to homeowners. However, after three days of continued hot weather, many customers may complain and opt out of the program. Newer technologies cycle the AC or allow the homeowner to reset the temperature in their home to respond to a grid event, thus not turning the AC off completely, but resetting the level of service or air conditioning duty cycle. Smart thermostat programs currently provide 1.2 GW of DR. There is a growing number of "Bring Your Own Thermostat" programs, where the utility provides incentives for smart thermostats that can communicate with the utility communication

Table 6. Existing Demand Response in the US in 2018.

Sector	End-Use	Demand (GW)
Residential	AC Switch	3.4
	Water Heaters	0.3
	Thermostats	1.2
	Behavioral	0.7
Commercial & Industrial	Automated	3.7
	Customer Initiated	7.2
Mass Market	Other	1.8
Total		18.3

Source: SEPA.[154]

technology to initiate systemwide DR events. Customers can more easily opt out of such systems if there is any inconvenience from participating in the event. Many of these smart thermostats also can pre-cool the home prior to the DR event, thus allowing the building mass to act as thermal storage and improve the comfort during the event.

Electric water heating is one of the well-known sources of flexible loads in homes. Utilities across the US have been exploring opportunities to use residential electric water heaters for grid services, and this area for DR is expected to grow. Water heaters provide a challenging end use, where the consideration between the value of energy efficiency and the value of demand response is being evaluated.[137] Most water heaters that provide demand response are electric resistance systems. These hot water storage systems can be charged prior to peak demand periods to minimize electric use during morning or evening peaks. There is a large research and testing effort in motion to provide heat pump water heaters that are both efficient and provide grid services. These, however, are more expensive than electric resistance systems, and require utility incentives or other policies to encourage their greater adoption.

The behavioral demand response category listed in Table 6 refers to the financial incentives that electric utilities offer customers to reduce their electric use during critical times. An important feature of these DR programs is the investment in smart meters that allow electric utilities to measure customer use at the hourly or more frequent time scale. The US has nearly 11 GW of DR in commercial and industrial programs, with 3.7 GW in the automated category. Both the automated and the customer-initiated manual DR includes HVAC and lighting end-use control, as well as industrial process control. The final category in Table 6 is the 1.8 MW in Mass Market DR, which includes both residential and small business customers. Small businesses are an underserved and hard to reach market segment, primarily because their owners are not as sophisticated as the large commercial and industrial customers, and their ability to invest in energy efficiency and DR technology is more challenging.

5.1.2. *Demand response automation*

Expansion of the market for advanced building control systems, and the ability to integrate them with utility DR systems, has been slowed by the lack of interoperability between proprietary products offered by competing vendors. Open standards would encourage competition and lower costs. The most prevalent open communication standard for automating demand response is OpenADR (open automated demand response). OpenADR was developed by LBNL and is managed through the OpenADR Alliance. OpenADR is an XML data model that uses a *client server* architecture, as shown in Fig. 45. The OpenADR architecture uses an automation *server*

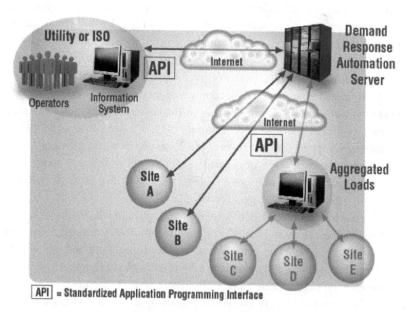

Fig. 45. The OpenADR client server architecture.

Note: API = Standardized Application Programming Interface.

to sends time-stamped signals from utilities to customers or aggregators to communicate with end-use loads. Customer control systems host the XML *client* to listen for the encrypted signals to ensure cybersecure communications. These continuous signals represent DR events, prices and grid needs. Buildings are then programmed to automatically change their operation to respond to the signal.

In 2018 OpenADR became on international standard sponsored by the International Electrotechnical Commission (IEC). The IEC approved the OpenADR 2.0b Profile Specification as a full IEC standard, to be known as IEC 62746-10-1 ED1. OpenADR is used around the world, with implementations in Europe, China, Japan and Korea. OpenADR is also required as part of the building energy code for commercial buildings in California, to reduce the cost to automate DR in new buildings.

5.2. *Advanced grid needs and grid interactive buildings*

Renewable energy deployment is accelerating around the world; initially driven by the imperative to reduce global warming, the technology is now cost competitive even without accounting for the climate externalities associated with fossil resources.[138] Increasing the use of renewables is supported by policy interventions like renewable portfolio standards that specify minimum grid mix levels. California, for example, has a renewable portfolio goal to achieve 33% of energy use by 2020 and 50% by 2030, established in Senate Bill 350.[139] As states and regions continue to add new renewable generation, a range of approaches are being used to reduce the cost of integrating these new resources, including changes to electricity transmission infrastructure, regional integration of grid management and deployment of distributed energy resources.

The renewables integration challenges for managing the electric grid, with significant contributions from solar, are exemplified by the "duck curve" described by the California System Operator.[140] Figure 46 shows how in 2017 the predicted "duck curve" had already been manifested in operations. The curve is characterized by a steep downward and upward ramping to manage morning and evening transitions in the availability of solar generation, and periods of "oversupply" in which renewable resources are curtailed to maintain grid stability. Renewables have significantly reduced and delayed the evening peak load, and have introduced new, steep ramps in the evening. Throughout the day, additional solar and wind power adds to short-run variability on the grid as well. From the period 2014 through 2017, as renewable contributions to the grid have grown, the

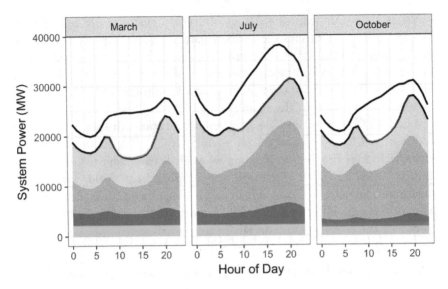

Fig. 46.　California ISO operations for three months in 2017 show the profile for average days, including generation from nuclear, hydroelectricity, in-state thermal and imports. Load and net load (load minus renewables) are shown with curtailment of renewable generation in red, layered next to the net load.

Source: Data from http://www.caiso.com/informed/Pages/ManagingOversupply.aspx.

average five-minute change in net load has increased by 28% (from an average magnitude of short-duration change of 102 MW to 131 MW over five-minute intervals).[g]

A recent study funded by the California Public Utilities Commission evaluated how DR can be used to address the challenges in the net load shape.[6] The net load shape is the difference between expected load and anticipated electricity production from renewable energy sources. One challenge in deploying renewable electricity systems is the need to raise public awareness of the new grid management dynamics that evolve with increased penetration of solar generation. For decades, the public message about the timing of demand (and the focus of time of use (TOU) rates) was rightfully focused on reducing loads on summer afternoons, when high air conditioning demand drove annual peak conditions. Customers were encouraged to use more electricity at night and in the early morning, and

---

[g]This outcome and others in this report related to CAISO operations between 2014 and 2017 are based on data obtained from the CAISO website in March 2018. http://www.caiso.com/informed/Pages/ManagingOversupply.aspx.

they responded appropriately, In addition, policy and technology initiatives supported measures that addressed that paradigm; for example, supporting ice storage systems that make ice at night and draw on the reservoir of "cold" in the daytime for cooling. However, with the deployment of solar generation, these needs have flipped. There is now a growing need to consume *more* electricity in mid-day periods on sunny days, when solar is generating, and the net load peak that matters for managing capacity has migrated into the early evening hours, when solar generation wanes. Reforms to TOU periods to align with these new needs have lagged behind this new paradigm, and the multi-year process for updating TOU rates is likely to continue to lag behind system conditions once new rates are deployed.

If dynamic load shaping is to be fully realized with day-to-day dispatch, there will need to be a fundamentally new compact between electricity suppliers and users who participate in DR programs. The conventional understanding that electricity prices are relatively constant, and that load use is disconnected from the conditions on the grid, with the occasional need for action (load shedding) needing to be replaced with a relationship of coordination for these customers. The delegation of authority to schedule and control loads from customers to automated systems will be critical for reducing DR transactions costs that would otherwise prevent many customers from participating, but such a delegation will require a degree of trust in the systems that are put in place. Cybersecurity, institutional responsibility for DR aggregators, and the perception of risk and benefits to customers will all be important factors as the DR market changes.

The Duck curve is seen in many regions of the US beyond California, including Hawaii and Vermont. The key issue is that there is a growing need to more strongly integrate energy supply with energy demand to promote clean, affordable and reliable energy systems.

6. Summary and Future Directions

This chapter has provided an overview of energy use in the buildings sector and an introduction to the technologies and practices to reduce energy use in buildings. Buildings are the largest energy using sector in the US and one of the fastest growing and largest energy loads around the world. Opportunities abound to reduce energy use in buildings.

Every topic covered in this chapter needs to be more fully researched, to explore how to improve energy efficiency in buildings and work toward low-cost and low-carbon systems. This research should include evaluating opportunities for the use of advanced materials in new insulation and windows, energy storage integrated with building heating and cooling, and information technology for control and optimization. We also need better research techniques to accelerate the deployment of energy efficiency, which might include creating new business models, exploring public private partnerships, collaborating with cities and local communities, and developing better decision making tools for business and consumers. Both technology and deployment research need better feedback loops on "what works" so we have a stronger community of learning and sharing. We need to put the new advances in social media and communications to work to ensure that better information is delivered to key stakeholders. Reducing energy use helps reduce the need for power plants, reduces greenhouse gas emissions, improves air quality and saves money.

As new technologies and opportunities to improve energy efficiency continue to develop, we are also facing an urgent need to update our metrics and cost effectiveness concepts to consider our changing electric grid. As the grid changes to include more variable renewables, we need to ensure that we invest in information, control, automation and measurement technologies that can provide both energy efficiency and flexible demand. These technologies will help us integrate end-use systems such as HVAC, lighting, dynamic façade and other end uses as whole-building assets. Controlling a building to integrate with PV and electric or thermal storage is becoming more important, as is the need for community scale, microgrid or district energy system integration. We need to understand how to track carbon intensity as a building performance metric. The GHG/kWh is becoming a dynamic, seasonal and hourly attribute of the grid and energy loads, and end-uses that can "charge" with cleaner electricity will become more valuable to achieve low-carbon buildings.

Acknowledgments

Much of the work described in this chapter was performed at the Lawrence Berkeley National Laboratory (LBNL) with funding from the US Department of Energy. The LBNL work was supported by the Assistant Secretary for Energy Efficiency and Renewable Energy, Building Technologies Office of the US Department of Energy under Contract No. DE- AC02-05CH11231.

References

1. Hoffman, I.M., C.A Goldman, S. Murphy, N. Mims Frick, G. Leventis, and L.C. Schwartz. 2018. "The cost of saving electricity through energy efficiency programs funded by utility customers: 2009–2015". LBNL Report. Lawrence Berkeley National Laboratory, Berkeley, California.
2. IEA, 2018a. "World energy balances (2018 edition)", International Energy Agency, 2018a. http://data.iea.org/payment/products/117-world-energy-balances-2018-edition-coming-soon.aspx.
3. United Nations, "World Population prospects: the 2017 revision, volume ii: demographic profiles", ST/ESA/SER.A/400, Department of Economic and Social Affairs, Population Division, United Nations, 2017.
4. Energy Information Administration, US Department of Energy. 2016. "2012 Commercial Buildings Energy Consumption Survey (CBECS)" https://www.eia.gov/consumption/commercial/.
5. Energy Information Administration, US Department of Energy. 2018a. "2015 Residential Energy Consumption Survey (RECS)" https://www.eia.gov/consumption/residential/.
6. Alstone, P., M.A. Piette, and P. Schwartz. 2018. "Integrating demand response and distributed resources in planning for large-scale renewable energy integration". *Proceedings of the 2018 ACEEE Summer Study on Energy Efficiency in Buildings*. Pacific Grove, California.
7. Rissman, J. and H. Kennan. 2013. "Low emissivity windows — case studies on the government's role in energy technology innovation". *American Energy Innovation Council*. https://energyinnovation.org/2013/03/07/aeic-technology-case-studies-released/.
8. Geller, H., and J. Thorne. 1999. "US Department of Energy's Office of Building Technologies: successful initiatives of the 1990s". *American Council for an Energy Efficient Economy* 2–8.
9. NFRC. 2019. "National Fenestration Rating Council". https://www.nfrc.org/. Accessed: May 2019.
10. EWC. 2019 "Efficient Window Collaborative". https://www.efficientwindows.org/. Accessed: May 2019.
11. LBNL. 2019a. "LBNL windows and daylighting group; software tools", Lawrence Berkeley National Laboratory. https://windows.lbl.gov/software. Accessed: May 2019.
12. AERC. 2019. "Attachment Energy Rating Council". https://aercenergyrating.org/. Accessed: May 2019.
13. LBNL. 2018. "LBNL Facades R&D Program", Lawrence Berkeley National Laboratory. https://facades.lbl.gov/demonstrations. Accessed: May 2019.
14. IEA. 2018b. "The future of cooling: opportunities for energy-efficient air conditioning, International Energy Agency. https://webstore.iea.org/the-future-of-cooling.

15. Chua, K. J., *et al.* 2013. "Achieving better energy-efficient air conditioning — a review of technologies and strategies". *Applied Energy* **104**, 87–104. http://dx.doi.org/10.1016/j.apenergy.2012.10.037.

16. Goetzler, W. 2007. "Variable refrigerant flow systems". *ASHRAE Journal* 24–31.

17. Amarnath, A., and M. Blatt. 2008. "Variable refrigerant flow: where, why, and how". *Engineered Systems* **25**(2), 54–60.

18. Amarnath, A., M. Blatt, and E.U. Consultant. 2008. "Variable refrigerant flow: an emerging air conditioner and heat pump technology evolution of the technology how does VRF work?" *ACEEE Summer Study on Energy Efficiency in Buildings.* Pacific Grove, California.

19. Liu, X. and T. Hong. 2010. "Comparison of energy efficiency between variable refrigerant flow systems and ground source heat pump systems". *Energy and Buildings* **42**(5), 584–589. http://dx.doi.org/10.1016/j.enbuild.2009.10.028.

20. Aynur, T.N. 2010. "Variable refrigerant flow systems: a review". *Energy and Buildings* **42**(7), 1106–1112. http://dx.doi.org/10.1016/j.enbuild.2010.01.024.

21. Aktacir, M.A., O. Büyükalaca, and T. Yilmaz. 2010. "A case study for influence of building thermal insulation on cooling load and air-conditioning system in the hot and humid regions". *Applied Energy* **87**(2), 599–607. http://dx.doi.org/10.1016/j.apenergy.2009.05.008.

22. Dyer, M. 2006. "Approaching 20 years of VRF in the UK". *Modern Building Services.* http://www.modbs.co.uk/news/fullstory.php/aid/2127/Approaching_20_years_of_VRF_in_the_UK.html.

23. Zhou, Y.P., *et al.* 2007. "Energy simulation in the variable refrigerant flow air-conditioning system under cooling conditions". *Energy and Buildings* **39**(2), 212–220.

24. Zhou, Y.P., *et al.* 2008. "Simulation and experimental validation of the variable-refrigerant-volume (VRV) air-conditioning system in EnergyPlus". *Energy and Buildings* **40**(6), 1041–1047.

25. Li, Y. M., J. Y. Wu, and S. Shiochi. 2010. "Experimental validation of the simulation module of the water-cooled variable refrigerant flow system under cooling operation". *Applied Energy* **87**(5), 1513–1521. http://dx.doi.org/10.1016/j.apenergy.2009.09.018.

26. Roth, K. W., *et al.* 2002. "Energy consumption characteristics of commercial building HVAC systems volume iii: energy savings potential". *Building Technologies Program* **III** (68370).

27. Zhu, Y., *et al.* 2014. "Control and energy simulation of variable refrigerant flow air conditioning system combined with outdoor air processing unit". *Applied Thermal Engineering* **64**(1–2), 385–395. http://dx.doi.org/10.1016/j.applthermaleng.2013.12.076.

28. Tu, Q., *et al.* 2010. "Heating control strategy for variable refrigerant flow air conditioning system with multi-module outdoor units". *Energy and Buildings* **42**(11), 2021–2027. http://dx.doi.org/10.1016/j.enbuild.2010.06.010.

29. Zhu, Y., *et al.* 2013. "Generic simulation model of multi-evaporator variable refrigerant flow air conditioning system for control analysis". *International Journal of Refrigeration* **36**(6), 1602–1615. http://dx.doi.org/10.1016/j.ijrefrig.2013.04.019.

30. Aynur, T.N., Y. Hwang, and R. Radermacher. 2006. "Field performance measurements of a Vrv Ac/Hp system". *International Refrigeration and Air Conditioning Conference 2005*, 1–8.

31. NEEP, "High performance rooftop units", Northeast Energy Efficiency Partnerships. https://neep.org/initiatives/high-efficiency-products/advanced-rooftop-units-artu.

32. US DOE, "Research and development roadmap for emerging HVAC technologies", Buildings Technologies Office, US Department of Energy, Washington, October 2014, p. 121.

33. ANSI/ASHRAE. 2012. "Panel heating and cooling", in *ASHRAE Handbook. HVAC Systems and Equipment* (American Society of Heating and Cooling).

34. US DOE, "Radiant heating", US Department of Energy. https://www.energy.gov/energysaver/home-heating-systems/radiant-heating.

35. Sattari, S., and B. Farhanieh. 2006. "A parametric study on radiant floor heating system performance". *Renewable Energy* **31**(10), 1617–1626.

36. Memon, R.A., S. Chirarattananon, and P. Vangtook. 2008. "Thermal comfort assessment and application of radiant cooling: a case study". *Building and Environment* **43**(7), 1185–1196.

37. Song, D., *et al.* 2008. "Performance evaluation of a radiant floor cooling system integrated with dehumidified ventilation". *Applied Thermal Engineering* **28**(11–12), 1299–1311.

38. US DOE, "Radiant cooling", US Department of Energy. https://www.energy.gov/energysaver/home-cooling-systems/radiant-cooling.

39. Tomlinson, J.J., C.K. Rice, and E. Baskin. 2006. "Integrated heat pumps for combined space conditioning and water heating". *8th IEA Heat Pump Conference, 2005: Global Advances in Heat Pump Technology, Applications, and Markets*, Las Vegas, Nevada, pp. 1–9.

40. Baxter, V.D., *et al.* 2017. "Air-source integrated heat pump development". Final report. US Department of Energy. https://info.ornl.gov/sites/publications/Files/Pub75419.pdf.

41. Dawson-Haggerty, S. 2018. "Works with Comfy: a new partner ecosystem for a better workplace experience". https://www.comfyapp.com/blog/announcing-works-with-comfy-a-new-partner-ecosystem/.

42. Kim, J., F. Bauman, P. Raftery, E. Arens, H. Zhang, G. Fierro, M. Andersen, and D. Culler. 2019. "Occupant comfort and behavior: high resolution data from a 6-month field study of personal comfort systems with 37 real office workers". *Building and Environment* **148**, 348–360. https://doi.org/10.1016/j.buildenv.2018.11.012.

43. Chen, Y., B. Raphael, and S.C. Sekhar. 2016. "Experimental and simulated energy performance of a personalized ventilation system with individual airflow control in a hot and humid climate". *Building and Environment* **96**, 283–292. http://dx.doi.org/10.1016/j.buildenv.2015.11.036.

44. Deng, Q., *et al.* 2017. "Human thermal sensation and comfort in a non-uniform environment with personalized heating". *Science of the Total Environment* **578**, 242–248. http://dx.doi.org/10.1016/j.scitotenv.2016.05.172.

45. Woerpel, H. 2018. "HVAC manufacturers adopt low-GWP refrigerants despite HFC ruling". https://www.achrnews.com/articles/136799-hvac-manufacturers-adopt-low-gwp-refrigerants-despite-hfc-ruling.

46. Brown, J.S. *et al.*, 2016. "Low-GWP refrigerants: guidance on use and basic competence requirements for contractors". *Science and Technology for the Built Environment* **22**(8), 1075–1076.

47. THERMA, "Low-GWP refrigerants and how it is affecting the HVAC industry", *Therma.com.* https://www.therma.com/low-gwp-refrigerants-and-how-it-is-affecting-the-hvac-industry/.

48. U.S. Department of Energy 2016. Solid State Lightening Research and Development Plan. U.S. Department of Energy, Building Technologies Office. Washington, D.C.

49. EIA, "Analysis and representation of miscellaneous electric loads in NEMS", Energy Information Administration, Washington, 2017. https://www.eia.gov/analysis/studies/demand/miscelectric/pdf/miscelectric.pdf.

50. US DOE, "Quadrennial technology review", US Department of Energy, Washington, 2015.

51. EIA, "EIA's residential energy survey now includes estimates for more than 20 new end uses", *Today in Energy*, Energy Information Administration, 5 June 2018a. https://www.eia.gov/todayinenergy/detail.php?id=36412&src=%E2%80%B9%20Consumption%20%20%20%20%20%20Residential%20Energy%20Consumption%20Survey%20(RECS)-b1#.

52. Rainer, L., A. Khandekar, and A. Meier. 2018. "Builder installed electric loads: the energy mortgage on a new house". *Proceedings of the ACEEE 2018 Summer Study on Energy Efficiency in Buildings.* Pacific Grove, California.

53. Gerber, Daniel L., Alan Meier, Robert Hosbach, and Richard Liou. 2018. "Zero standby solutions with optical energy harvesting from a laser pointer". *Electronics* **7**, 292. https://doi.org/10.3390/electronics7110292.

54. Gaetani, I., P.-J Hoes, and J.L.M. Hensen. 2016. "Occupant behavior in building energy simulation: towards a fit-for-purpose modeling strategy". *Energy and Buildings* **121**, 188–204.

55. Yoshino, H., T. Hong, and N. Nord. 2017. "IEA EBC annex 53: total energy use in buildings — analysis and evaluation methods". *Energy and Buildings* **152**, 124–136. http://dx.doi.org/10.1016/j.enbuild.2017.07.038.

56. Chen, Y., X. Luo, and T. Hong. 2016. "An agent-based occupancy simulator for building performance simulation". *ASHRAE Annual Conference.*
57. Marmaras, J.M. 2014. "Measurement and verification — retro-commissioning of a LEED gold rated building through means of an energy model: are aggressive energy simulation models reliable?"
58. Morari, M., and J. Lee. 1997. "Model predictive control: past, present and future". *Computers & Chemical Engineering* **23**(4–5), 667–682.
59. Salakij, S., *et al.* 2016. "Model-based predictive control for building energy management. I: Energy modeling and optimal control". *Energy and Buildings* **133**, 345–358.
60. Crawley, D.B., *et al.* 2001. "EnergyPlus: creating a new-generation building energy simulation program". *Energy and Buildings* **33**(4), 319–331.
61. Yan, D., *et al.* 2008. "DeST — an integrated building simulation toolkit, part I: fundamentals". *Building Simulation* **1**(2), 95–110. http://link.springer.com/10.1007/s12273-008-8118-8%5Cnhttp://link.springer.com/article/10.1007/s12273-008-8118-8%5Cnhttp://link.springer.com/content/pdf/10.1007/s12273-008-8118-8.pdf%5Cnhttps://www.researchgate.net/publication/226594932_DeST__An_integrated_buil.
62. ESRU, "The ESP-r system for building energy simulation: user guide version 10 series", Energy Systems Research Unit, University of Strathclyde, Glasgow, Scotland, 2003.
63. Kalamees, T., 2004. "IDA ICE: the simulation tool for making the whole building energy and HAM analysis". *Annex 2004 MOIST-ENG, Working Meeting,* Zurich, Switzerland.
64. Klein, S.A. 1988. "TRNSYS — a transient system simulation program". University of Wisconsin-Madison, Engineering Experiment Station.
65. Hong, T., *et al.* 2016b. "An occupant behavior modeling tool for co-simulation". *Energy and Buildings* **117**, 272–281.
66. Hong, T., *et al.* 2015. "Commercial building energy saver: an energy retrofit analysis toolkit". *Applied Energy* **159**, 298–309. http://linkinghub.elsevier.com/retrieve/pii/S0306261915010703.
67. Chen, Y., T. Hong, and M.A. Piette. 2017. "Automatic generation and simulation of urban building energy models based on city datasets for city-scale building retrofit analysis". *Applied Energy* **205**, 323–335. http://dx.doi.org/10.1016/j.apenergy.2017.07.128.
68. Hong, T., *et al.* 2016a. "CityBES: a web-based platform to support city-scale building energy efficiency". *5th International Urban Computing Workshop.* San Francisco.
69. Lubliner, M. and R. Kunkle. 2015. "A case study of residential new construction ductless heat pump performance and cost effectiveness". Tacoma Power report. Washington State University.

70. Lubliner, M., M. Spencer, L. Howard, D. Hales, R. Kunkle, A. Gordon, and M. Spencer. 2016. "Performance and costs of ductless heat pumps in marine-climate high-performance homes — habitat for humanity the woods". Report for US DOE Building America. Washington State University.

71. Larson, B. 2017. "Laboratory assessment of EcoRuno CO_2 air-to-water heat pump". Washington State University. http://www.energy.wsu.edu/ Documents/Sanden EcoRuno Lab Test Report.pdf. http://www.energy.wsu. edu/Documents/Sanden EcoRuno Lab Test Report.pdf.

72. International Code Council, "International energy conservation code", 2018.

73. Chan, W.R., J. Joh, and M.H. Sherman. 2013. "Analysis of air leakage measurements of US houses". *Energy and Buildings* **66**, 616–625.

74. NRDC, "Home idle load: devices wasting huge amounts of electricity when not in active use", NRDC Issue Paper, Natural Resources Defense Council, 2015.

75. Taylor, N., J. Searcy, and P. Jones. 2016. "Multifamily energy-efficiency retrofit programs: a Florida case study". *Energy Efficiency* **9**, 385–400. DOI 10.1007/s12053-015-9367-x.

76. RDH, "Energy consumption and conservation in mid- and high-rise residential buildings in British Columbia", Report for Canada Mortgage and Housing Corporation, RDH Building Engineering, 2012.

77. Bardacke, T., and W. Wells. 2012. "Affordable multifamily zero energy new homes". CEC-500-2012-052. Global Green USA. California Energy Commission.

78. Sherman, M.H., and I.S. Walker. 2011. "Meeting residential ventilation standards through dynamic control of ventilation systems". *Energy and Buildings* **43**, 1904–1912. LBNL-4591E. Elsevier.

79. Guyot, G., M.H. Sherman, and I.S. Walker. 2018. "Smart ventilation energy and indoor air quality performance in residential buildings: a review". *Energy and Buildings* **163**, 416–430. https://doi.org/10.1016/j.enbuild.2017.12.051.

80. Less, B.D., N. Mullen, B. Singer, and I.S. Walker. 2015. "Indoor air quality in 24 California residences designed as high performance homes". *Science and Technology for the Built Environment* **21**, 14–24. Doi: 10.1080/10789669. 2014.961850. LBNL-6937E.

81. Breysse, J., D.E. Jacobs, W. Weber, S. Dixon, C. Kawecki, S. Aceti, and J. Lopez. 2011. "Health outcomes and green renovation of affordable housing". *Public Health Reports* **126**, 64–75.

82. Jacobs, D.E. 2013. "Health outcomes of green and energy-efficient housing". Presented at the Lead & Environmental Hazards Association, Peoria.

83. Leech, J. A., M. Raizenne, and J. Gusdorf. 2004. "Health in occupants of energy efficient new homes". *Indoor Air* **14**(3), 169–173. doi:10.1111/j.1600-0668.2004.00212.x.

84. Kovesi, T., *et al.* 2009. "Heat recovery ventilators prevent respiratory disorders in Inuit children". *Indoor Air* **19**(6), 489–499. doi:10.1111/j.1600-0668.2009.00615.x.

85. Noris, F., G. Adamkiewicz, W.W. Delp, T. Hotchi, M. Russell, B.C. Singer, M. Speard, K. Vermeer, and W.J. Fisk. 2013. "Indoor environmental quality benefits of apartment energy retrofits". *Building and Environment* **68**, 170–178. doi:10.1016/j.buildenv.2013.07.003.

86. Weichenthal, S., G. Mallach, R. Kulka, A. Black, A. Wheeler, H. You, and D. Sharp. 2013. "A randomized double-blind crossover study of indoor air filtration and acute changes in cardiorespiratory health in a First Nations community". *Indoor Air* **23**(3), 175–184. doi:10.1111/ina.12019.

87. Coulter, J., B. Davis, C. Dastur, M. Malkin-Weber, and T. Dixon. 2007. "Liabilities of vented crawl spaces and their impacts on indoor air quality in southeastern US Homes". *Clima 2007 WellBeing Indoors*.

88. Emmerich, S.J., J.E. Gorfain, M. Huang, and C. Howard-Reed. 2003. "Air and pollutant transport from attached garages to residential living spaces". *NISTIR* 7072, **25**.

89. Logue, J.M., P.N. Price, M.H. Sherman, and B.C. Singer. 2012. "A method to estimate the chronic health impact of air pollutants in US residences". *Environmental Health Perspectives* **120**(2), 216–222.

90. Singer B., W. Delp, D. Black, and I. Walker. 2016. "Measured performance of filtration and ventilation systems for fine and ultrafine particles and ozone in a modern California house". *Indoor Air* **27**, 780–790. doi: 10.1111/ina.12359. LBNL 1006961.

91. Hult, E., H. Willem, P. Price, T. Hotchi, M. Russel, and B. Singer. 2014. "Formaldehyde and acetaldehyde exposure mitigation in US Residences: in-home measurements of ventilation control and source control". *Indoor Air* **25**. doi.org/10.1111/ina.12160.

92. Singer B., and W. Delp. 2018. "Response of consumer and research grade indoor air quality monitors to residential sources of fine particles". *Indoor Air* **28**(4), 624–639. DOI: 10.1111/ina.12463.

93. Chan, W.R., Y.-S. Kim, B.B. Less, B.C. Singer, and I.S. Walker. 2018. "Ventilation and indoor air quality in new California Homes with gas appliances and mechanical ventilation". California Energy Commission.

94. Less, B. 2012. "Indoor air quality in 24 California residences designed as high performance green homes". University of California, Berkeley, California. http://escholarship.org/uc/item/25x5j8w6.

95. Hill, D. 1998. "Field survey of heat recovery ventilation systems". Technical Series No. 96-215. Research Division, Canada Mortgage and Housing Corporation, Ottawa, Ontario. http://publications.gc.ca/collections/collection_2011/schl-cmhc/nh18-1/NH18-1-90-1998-eng.pdf.

96. BTO, "Building Technology Office", 2019. https://www.energy.gov/eere/buildings/appliance-and-equipment-standards-program.

97. EIA, "LED lightbulbs keep improving in efficiency and quality", *Today in Energy*, Energy Information Administration, 4 November 2014. https://www.eia.gov/todayinenergy/detail.php?id=18671. Accessed: July 2019.

98. Hart, R., D. Morehouse, W. Price, J. Taylor, H. Reichmuth, and M. Cherniack. 2008. "Up on the roof: from the past to the future". *2008 ACEEE Summer Study on Energy Efficiency in Buildings.* https://aceee.org/files/proceedings/2008/data/papers/3_295.pdf.

99. Fernandez, N., Y. Xie, S. Katipamula, M. Zhao, W. Wang, and C. Corbin. 2017. "Impacts of commercial building controls on energy savings and peak load reduction". PNNL-25985. Prepared for the US Department of Energy.

100. Kramer, H., Lin, G., Curtin, C., Crowe, E., Granderson, J. 2020. "Building analytics and monitoring-based commissioning: industry practice, costs, and savings". *Energy Efficiency* **13**, 537–549. DOI: https://doi.org/10.1007/s12053-019-09790-2.

101. ENERGY STAR, "ENERGY STAR guidelines for energy management", ENERGY STAR, 2016. https://www.energystar.gov/buildings/tools-and-resources/energy-star-guidelines-energy-management. Accessed: July 2019.

102. Mathew, P., D. Sartor, O. van Geet, S. Reilly. "Rating energy efficiency and sustainability in laboratories: results and lessons from the Labs21 program", *Proceedings of the 2004 ACEEE Summer Study of Energy Efficiency in Buildings.* Pacific Grove, California.

103. ENERGY STAR, "ENERGY STAR portfolio manager", ENERGY STAR, 2019. https://portfoliomanager.energystar.gov/pm/login.html. Accessed: February 2019.

104. ENERGY STAR, "Portfolio Manager Technical Reference: ENERGY STAR score", ENERGY STAR, 2018. https://www.energystar.gov/buildings/tools-and-resources/portfolio-manager-technical-reference-energy-star-score. Accessed: February 2019.

105. EIA, "Commercial building energy consumption survey", Energy Information Administration, 2019. https://www.eia.gov/consumption/commercial/. Accessed: February 2019.

106. ASHRAE, "Building EQ", ASHRAE, 2019. https://www.ashrae.org/technical-resources/building-eq. Accessed: Feruary 2019.

107. PNNL, "Asset score", Pacific Northwest National Laboratory, 2019. https://buildingenergyscore.energy.gov/. Accessed: February 2019.

108. LBNL, "DOE building performance database", Lawrence Berkeley National Laboratory, 2019b. https://bpd.lbl.gov/. Accessed: February 2019.

109. I2SL, "Laboratory Benchmarking Tool", International Institute for Sustainable Laboratories, 2019. https://lbt.i2sl.org/. Accessed: February 2019.

110. ASHRAE, "Standard for commercial building energy audits", Standard 211-2018, ASHRAE, 2018.

111. Zhu, C., A. White, P. Mathew, J. Deason, and P. Coleman. 2018. "Raising the rent premium: moving green building research beyond certifications and rent". *ACEEE Summer Study on Energy Efficiency in Buildings.* American Council for an Energy Efficient Economy.

112. Smith, K.R., S. Mehta, and M. Maeusezahl-Feuz. 2004. "Indoor smoke from household solid fuels". in *Comparative Quantification of Health Risks: Global and Regional Burden of Disease due to Selected Major Risk Factors.* WHO, Geneva, Switzerland. 1435–1493.

113. Gadgil, A.J., A. Sosler, and D. Stein. 2013. "Stove solutions: improving health, safety, and the environment in Darfur with fuel-efficient cookstoves". *The Solutions Journal* 4, 54–64.

114. Zhou, N., N. Khanna, W. Feng, J. Ke, and M. Levine. 2018. "Scenarios of energy efficiency and CO_2 emissions reduction potential in the buildings sector in China to year 2050". *Nature Energy* 3, 978–984. https://doi.org/10.1038/s41560-018-0253-6.

115. Tsinghua University Building Energy Research Center, *China Building Energy Efficiency Development and Research Annual Report 2011* (Beijing: China Construction Industry Press, 2011).

116. ERI, LBNL and RMI, "Reinventing fire: China — a roadmap for China's revolution in energy consumption and production to 2050", Executive Summary, Energy Research Institute, Lawrence Berkeley National Laboratory and Rocky Mountain Institute, 2016.

117. ERI, LBNL and RMI, "Reinventing fire: China — a roadmap for China's revolution in energy consumption and production to 2050", Executive Summary, Energy Research Institute, Lawrence Berkeley National Laboratory and Rocky Mountain Institute, 2016.

118. National Bureau of Statistics of China (NBS), *China 2017 Statistical Yearbook* (China Statistics Press, Beijing, 2018).

119. ERI, LBNL and RMI, 2016.

120. ERI, LBNL and RMI, 2016.

121. Zheng, X., C. Wei, P. Qin, J. Guo Y. Yu, F. Song, and Z. Chen. 2014. "Characteristics of residential energy consumption in China: findings from a household survey". *Energy Policy* 75, 126–135. https://doi.org/10.1016/j.enpol.2014.07.016.

122. Zhou, N., D. Fridley, N. Zheng Khanna, J. Ke, M. McNeil, and M. Levine. 2013. "China's energy and emissions outlook to 2050: perspectives from bottom-up energy end-use model". *Energy Policy* 53, 51–62. https://doi.org/10.1016/j.enpol.2012.09.065.

123. Zhou, N., N. Khanna, W. Feng, J. Ke, and M. Levine. 2018. "Scenarios of energy efficiency and CO_2 emissions reduction potential in the buildings sector in China to year 2050". *Nature Energy* **3**, 978–984. https://doi.org/10.1038/s41560-018-0253-6.

124. Javalgi, R.R.G., and D.A. Grossman. 2016. "Aspirations and entrepreneurial motivations of middle-class consumers in emerging markets: the case of India". *International Business Review* **25**(3): 657–667. https://doi.org/10.1016/j.ibusrev.2015.10.008.

Energy Efficiency

125. Bhattacharyya, S.C. 2015. "Influence of India's transformation on residential energy demand". *Applied Energy* **143**, 228–237. https://doi.org/10.1016/j.apenergy.2015.01.048.

126. de la Rue du Can, S., A. Khandekar, N. Abhyankar, N. Khanna, D. Fridley, N. Zhou, and A.A. Phadke. 2019. "Modeling India's energy future using a bottom-up approach". *Applied Energy* **238**, 1108–1125. 10.1016/j.apenergy.2019.01.065.

127. Zhou, N., D. Fridley, N. Zheng Khanna, J. Ke, M. McNeil, and M. Levine. 2013. "China's energy and emissions outlook to 2050: perspectives from bottom-up energy end-use model". *Energy Policy* **53**, 51–62. https://doi.org/10.1016/j.enpol.2012.09.065.

128. Yuan, X., X. Zhang, J. Liang, Q. Wang, and J. Zuo. 2017. "The development of building energy conservation in China: A review and critical assessment from the perspective of policy and institutional systems". *Sustainability* **9**, 1654–1676. doi:10.3390/su9091654.

129. N Khanna, N., J. Romankiewicz, W. Feng, N. Zhou, and Q. Ye. 2014. "Comparative policy study for green buildings in US and China". LBNL Report 6609-E. Lawrence Berkeley National Laboratory, Berkeley, California. https://china.lbl.gov/sites/all/files/green_buildings_policy_comparison.pdf.

130. NRDC, *"Dispersed coal report 2017"*, NRDC, 2017. In Chinese.

131. Abhyankar, N., N. Shah, V.E. Letschert, and A.A. Phadke. 2017. "Assessing the cost-effective energy saving potential from top-10 appliances in India". *9th International Conference on Energy Efficiency in Domestic Appliances and Lighting (EEDAL)*. http://eta-publications.lbl.gov/sites/default/files/india_appliance_ee_potential_eedal_conference_paper_0.pdf.

132. Kumar, S., N. Kumar, K. Cherail, S. Setty, N. Yadav, and A. Goenka. 2017. "Transforming the energy services sector in India–towards a billion dollar ESCO market". Alliance for an Energy Efficient Economy.

133. AEEE, "Alliance for an Energy Efficiency Economy", ECBC Implementation, Alliance for an Energy Efficiency Economy, 2019. http://www.aeee.in/projects/ecbc-implementation/.

134. Shah, N., M. Wei, V. Letschert, and A. Phadke. 2015. "Benefits of leapfrogging to superefficiency and low global warming potential refrigerants in room air conditioning". Lawrence Berkeley National Laboratory. http://eta-publications.lbl.gov/sites/default/files/lbnl-1003671.pdf.

135. Park, W.Y., N. Shah, and B. Gerke. 2017. "Assessment of commercially available energy-efficient room air conditioners including models with low global warming potential (GWP) refrigerants". LBNL-2001047. Lawrence Berkeley National Laboratory. http://eta-publications.lbl.gov/sites/default/files/assessment_of_racs_lbnl-_2001047.pdf.

136. de la Rue du Can, S., G. Leventis, A. Phadke, and A. Gopal. 2014. "Design of incentive programs for accelerating penetration of energy-efficient

appliances". *Energy Policy* **72**, 56–66. https://doi.org/10.1016/j. enpol.2014.04.035.

137. Podorson, D. 2016. "Grid interactive water heaters, how water heaters have evolved to a grid scale energy storage medium". *Proceedings of the 2016 ACEEE Summer Study on Energy Efficiency in Buildings*, Pacific Grove, California.

138. Lazard, "Levelized Cost of Energy 2017", Lazard, 2017. https://www.lazard.com/perspective/levelized-cost-of-energy-2017/.

139. California State Senate, "Clean Energy & Pollution Reduction Act. SB350", California State Senate, 2015. http://www.energy.ca.gov/sb350/.

140. CAISO, "What the Duck curve tells us about managing a green grid", CAISO, 2016. https://www.caiso.com/Documents/FlexibleResourcesHelp Renewables_FastFacts.pdf.

141. Lee, E.S. *et al.* 2020. "High-Performance Integrated Window and Façade Solutions for California". California Energy Commission report CECF-500-2020-001.

142. Karunakaran, R., S. Iniyan, and R. Goic. 2010. "Energy efficient fuzzy based combined variable refrigerant volume and variable air volume air conditioning system for buildings". *Applied Energy* **87**(4), 1158–1175. http://dx.doi.org/10.1016/j.apenergy.2009.08.013.

143. US Department of Energy (DOE). 2009. "Building technologies program: building energy software tools directory". http://apps1.eere.energy.gov/buildings/tools_directory/.

144. ANSI/ASHRAE. 2017. "Method of test for the evaluation of building energy analysis computer programs". http://www.techstreet.com/ashrae/products/1888937.

145. Walker, I.S., M.H. Sherman, and B.D. Less. 2014. "Houses are dumb without smart ventilation". *Proceedings of the ACEEE Summer Study 2014* LBNL 6747E. Washington.

146. Less, B.D., and I.S. Walker. 2014. "A meta-analysis of single family deep energy retrofit performance in the US". LBNL-6601E.

147. EIA. 2018b. "EIA's residential energy survey now includes estimates for more than 20 new end uses". *Today in Energy*, 5 June 2018. https://www.eia.gov/todayinenergy/detail.php?id=36412&src=%E2%80%B9%20Consumption%20%20%20%20%20%20Residential%20Energy%20Consumption%20Survey%20(RECS)-b1#.

148. Arasteh, D., *et al.* 2006. "Zero energy windows". *Proceedings of the 2006 ACEEE Summer Study on Energy Efficiency in Buildings*. Pacific Grove, California.

149. Ducker. 2018. Data supplied by Nick Limb, Ducker Research.

150. UPONOR. 2016. "Radiant cooling solutions improve DOAS".

151. Sola, A., *et al.* 2018. "Simulation tools to build urban-scale energy models: a review". *Energies* **11**(12), 3269. http://www.mdpi.com/1996-1073/11/12/3269.

152. IMT. 2019a. "US building benchmarking policy landscape". Institute for Market Transformation. 2019. https://www.buildingrating.org/graphic/us-building-benchmarking-policy-landscape. Accessed: February 2019.

153. IMT. 2019b. "US building area covered annually". Institute for Market Transformation. 2019. https://www.buildingrating.org/graphic/us-building-area-covered-annually. Accessed: February 2019.

154. Smart Electric Power Alliance (SEPA). 2018. "Utility demand response market snapshot".

Chapter 3

Industrial Energy Efficiency

Henry Kelly

*Institute for Sustainable Energy, Boston University, Boston,
MA 02215, USA*

henry.c.kelly@gmail.com

1. Introduction

Everyone has a basic idea of how buildings and vehicles use energy, but industrial energy efficiency involves a vast array of products and processes — famously including both computer chips and potato chips. The sector uses every form of energy, including coal, oil, gas, biomass, utility and distributed electricity, and a bewildering array of production processes like farming, construction, production of fuels and chemicals, production of steel and other metals, production of machinery and many other products. Each of these industrial categories uses energy in unique ways. The array of innovations needed to increase energy productivity is diverse and governed by numerous regulations and other policies. This diversity, coupled with the fact that many industrial processes are opaque to non-specialists, has made industrial energy analysis exceptionally difficult. There is a compelling need to find a way to cut through this complexity and develop a coherent set of research priorities and policy incentives that can match the scale and urgency of the problem. The only policies likely to produce incentives in all categories would be strategies for putting a price on greenhouse gases released into the atmosphere, and strategies for encouraging research, demonstration and deployment, including federally sponsored research.

While there are reasons to doubt the accuracy of some of the data for non-Organization for Economic Co-operation and Development (OECD) nations, Fig. 1[1] provides a rough view of how the world's industry uses

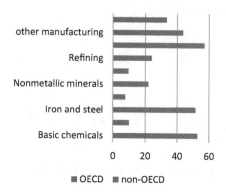

Fig. 1.   Global industrial energy use (EJ).

energy. While a few industries are large energy consumers, the bulk of industrial use is scattered across numerous categories. Whenever a large fraction of a sector's energy use appears as "other", the challenge is clear.

The mix of products depends heavily on the state of a nation's economic development. Advanced economies generate most of their income from services and products that have a high ratio of value to weight (i.e., cell phones), which allows them to increase both the efficiency of energy use and the value (added per kg) of other raw materials, while emerging economies that are building basic infrastructure like new highways and cities are producing vast amounts of steel and concrete. This chapter will explore the entire spectrum.

Defining efficiency in industry, however, is much more difficult than defining efficient buildings or transportation systems:

- Some processes, especially cement, result in the release of carbon dioxide through energy used to power the process and emissions from the chemical process itself.
- Changes in industrial processes can result in energy savings not measured in the production facility. Production of lightweight materials, for example, can lead to improvements in vehicle efficiencies, and innovative manufacturing methods can produce complex shapes that greatly improve performance per unit weight.
- Fossil fuels are the major raw materials for refineries and petrochemical industries that, together, make up over 40% of all industrial uses of energy (including the fuel used as feedstocks).

- Energy use can be greatly reduced if the products produced are recycled — a process that is only partly under the control of industry.

The discussion that follows will take a flexible view of "energy efficiency" and explore ways to reduce processing energy and the use of fossil fuel feedstocks and net release of carbon dioxide.

Manufacturing energy use is likely to be reshaped by policies that stand a good chance of meeting contemporary goals in climate change. The petrochemical manufacturing sector will see its markets all but vanish since most of its products are consumed in vehicles and applications where the carbon dioxide released can be captured. Markets will instead rely on production from biomass, zero-carbon electricity, hydrogen, and other low-carbon energy carriers. Cement and steel together are responsible for about a third of industrial energy use and most existing production methods rely on very high-temperature processes. Biofuels or fuels manufactured from renewable resources can substitute for the fossil fuels used now, or new materials may be used as substitutes for conventional cement and steel. Basic chemicals, made largely from fossil fuels, will also need to find new feedstocks and new chemistries, possibly mimicking biological processes. Taken together, estimates suggest that known technologies can by 2040 at least double the value created by a unit of industrial energy.[2] However, meeting climate goals will require driving greenhouse gas emissions to zero (or to negative values with systems that can remove carbon dioxide from the atmosphere).[3]

While the challenges are complex, an enormous array of innovative concepts are being developed that can address them. Some are straightforward improvements to existing production methods, while others rely on radically new approaches. Low cost sensors, communication and analytic tools will drive efficiency in the design and operation of industrial systems. New computational tools can also improve the design of products by minimizing material use and be used to design new materials that include composites, carbon fibers and bio-inspired materials. Novel methods for transferring heat and moisture (i.e., solid-state heat pumps, ultrasound, microwave and membranes) can replace fuel-based methods. Carbon fiber and composites and new process chemistry based in large parts on computer simulations may open many new horizons. A new understanding of biological materials and biological synthesis may revolutionize chemical production.

The rate at which these innovations are developed and introduced into the economy will depend heavily on public policy, which can support research and development or create incentives for efficient production methods with low greenhouse gas emissions. Only a few of these concepts can be covered in the discussion that follows, but we will explore some of the most important ones.

1.1 *How industry uses energy*

Industrial energy use is difficult to document since it can be measured in many different ways: by industry (e.g., food processing, steel), by process (i.e., drying, high-temperature firing and electro-chemical), or by fuel type (i.e., fuels and electricity). Figure 2 provides a window into how the US industry uses energy. Nearly half of the total primary energy consumed by industry is used for process heating. This includes the losses from electric generation and steam production.

The category "process heat" covers an enormous range of applications and process temperatures. Figure 3[4] shows a breakdown by industry type

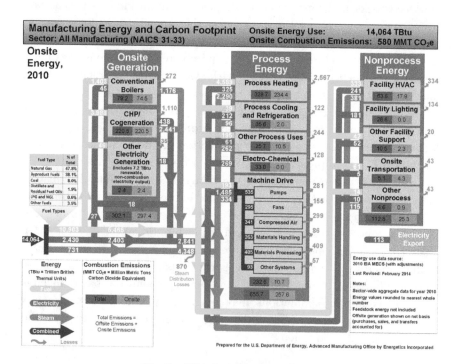

Fig. 2. US industrial energy flows.

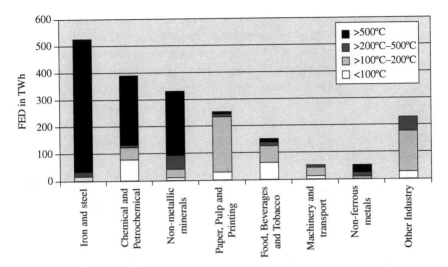

Fig. 3.   Industrial process heat demand for Europe (EU28+3).

Source: M. Rehfeldt, T. Fleiter and F. Toro, "A bottom-up estimation of the heating and cooling demand in European industry", *Energy Efficiency* **11** (2018) 1057. https://www.researchgate.net/publication/312173884_A_bottom-up_estimation_of_heating_and_cooling_demand_in_the_European_industry.

and temperature. About half of industrial process energy is used in producing metals, chemicals, cement and other "non-metallic" materials. These industries dominate demand for very high temperature processes.

Machine drive is responsible for about a quarter of the total industrial energy demand and nearly 60% of the electricity consumed. (The chart does not show the energy used to generate the purchased electricity. On average about two-thirds of the fossil energy burned to generate electricity is lost as waste heat.) About 10% of all industrial energy and nearly 20% of industrial electricity is used to heat, air condition, light and provide other services for the facilities. This of course could change if "lights out" factories employing robots become common. Building energy efficiency technologies must be a part of industrial energy efficiency strategies.

2.   Process Improvements

2.1.   *Electrification digitization and control*

The net efficiency of an industrial process requires both efficient devices and optimized design and operation of the integrated systems that use these devices. Integration involves "machine-to-plant-to-enterprise-to-supply-chain aspects of sensing", instrumentation, monitoring, control, and optimization as well as hardware and software platforms for industrial automation.[5] These systems are designed to continuously monitor performance at many different points in the operation, immediately identifying inefficiencies and anomalies, and promptly taking appropriate remedial actions. This can, for example, result in systems that do not repeatedly reheat material during processing, ensure efficient mass and optimize heat transfer and reaction kinetics.[6]

The net savings resulting from these modern systems result largely from integrated control over two critical technologies: machine drive (mostly responsible for the 25% of industrial energy), and management of waste heat. Figure 2 suggests that 2.6 quads are lost to waste heat by industrial processes, not including the waste created by generating the fuels and electricity that power the processes. Recent analysis suggests that improved process integration can reduce by half the cost of producing some complex composite materials and the energy needed by a factor of five.[7]

The ability to control production processes is greatly facilitated by sharply falling prices and rapid improvements in sensors, communication systems (internet of things), tools for managing vast amounts of data, solid-state controls (using a new generation of wide band-gap semiconductors), and improved analytic tools. Control is also facilitated by the shift to electric production processes. Robots, heat pumps, microwaves and other electric devices allow much tighter control over how energy is used.[8]

2.2. *Machine drive*

Machine drive systems can be improved by increasing the efficiency of the motors, the design of the equipment they operate, and the way these systems are controlled. The last two categories prove to be the most important. New materials and designs can improve the efficiency of motors, but the benefits will accrue mostly in smaller motors; large motors already operate at efficiencies greater than 95%.[9] A recent analysis of optimized systems of motor drives showed that 13% of the savings resulted from better motors. The use of variable speed drives accounted for 25% of the savings, and 62% of the savings resulted from improvements to the devices they operate (i.e., pumps, fans, compressors and blowers).[10] Savings result from improved designs and operational strategies, and from the use of lightweight materials. Lightweight components can cut energy use in robotics and other devices with parts that must accelerate and decelerate repeatedly.

An example of what can be achieved can be found in processes that involve controlling the flow of liquids or gases. Many existing systems are sized to meet peak demands and waste energy when loads are low. Control is often achieved by throttling flows with valves while operating the pumps and fans at constant speeds. Using variable speed drives that allow motors to closely match power demands without sacrificing efficiency can make large improvements.[11] The variable speed systems can, in turn, be improved using low cost, high-capacity wide bandgap semiconductors.[12]

2.3. *Waste heat*

The heat wasted in industrial processes can be reduced by improved process control (avoiding repeated heating and cooling of materials during processing), better insulation and heat containment, and systems that generate electricity by capturing otherwise wasted heat. It seems possible that 2–5 EJ of energy could be saved in the US industry by using existing technologies.[13] There are numerous practical limits on how much waste heat can be reduced. New types of heat exchangers that can resist fouling and operate at extreme temperatures would help. Thermoelectric devices could play a key role in waste heat recovery if costs and performance can be improved.

By using co-generation systems that produce electricity and steam, system-level waste can be reduced. However, climate policies will reduce options for the fuel used in such systems. Fossil fuels can only be used if the carbon dioxide is captured and sequestered — a technology likely to be feasible only for very large facilities. Biomass or a renewable fuel manufactured using renewable resources could also be used.

2.4. *Process revolutions*

While the efficiency of existing production methods can be greatly improved, a variety of dramatically new approaches to manufacturing are beginning to enter the market. These include replacing traditional subtractive production with additive production, the use of heat pumps for process heat to replace conventional fuel-powered systems, the use of membranes and other innovative techniques for separations that would

otherwise use energy-intensive, distillation, and mimicking the way biological systems produce a complex array of products from simple materials
at room temperature.

2.4.1. *Additive manufacturing*

Additive manufacturing produces products by repeatedly adding small
amounts of material to the part being produced rather than a "subtractive" process where material is removed by milling or other means. This
can improve efficiency in three ways:

- *It can reduce the use of raw materials.* Parts can be designed to use
  material only where they are needed to resist stress or strain or provide
  other functions. In many cases, these designs involve complex shapes
  and interior cavities that simply cannot be manufactured using conventional methods. Material savings of up to 90% are possible.[14]
- *The production process itself can be more efficient.*
- *The optimized, lightweight parts produced can increase the efficiency of
  vehicles, robots and other devices using them.*

Quantifying these benefits can be difficult. An analysis of parts used
in cars, trucks and aircrafts showed that for cars and trucks, more than
90% of the difference between subtractive and additive manufacturing
resulted from energy savings in the production process, while 60–70% of
the savings for aircrafts resulted from fuel used by them.[15] In the latter,
savings from a reduced use of materials were minimal. A similar study of
the production of an aluminum diesel engine pump housing showed that
savings were roughly divided between the reduced use of raw materials
and the use of fuel by the final product.[16]

2.4.2. *Novel approaches to heating, drying and separation*

Most of the low temperature heating processes shown in Fig. 3[17] use fuels,
primarily natural gas. Electric heat pumps can now provide the heat
needed in many of these applications. The new technologies offer high
efficiencies even at part load and can operate in a variety of temperature ranges.[18] The most attractive application areas are found where
waste heat can be captured and reheated to production temperatures, or

where heating, drying and refrigeration are needed. An example is the pasteurization process where milk or other products are raised to the temperature needed for pasteurization and then refrigerated for shipment.[19] Here, energy savings as high as 80% are possible.[20]

Separations, including water-removal, are traditionally made with fuel-powered distillation processes in many areas, including pulp and paper, desalination, waste concentration, and food processing. Petrochemical and other chemical productions use over half of the energy used for industrial separations, much of it at high temperature.[21] Membrane and advanced filtration technologies can offer dramatic efficiency improvements.[22] Membranes for separating products at low temperatures are often made of organic materials and high-temperature membranes are typically made from metals such as palladium.

Several other new approaches to remove water in applications are being explored. These range from clothes drying to industrial-scale driers in pulp and paper production. Other options include ultra sound, far-infrared, microwave and "super critical assisted" drying.[23]

2.4.3. *Bio-inspired manufacturing*

The fact that biological systems can produce an enormous variety of products efficiently from abundant raw materials at room temperature creates an intriguing existence proof for industrial production designers.[24] It has proven extremely difficult to mimic biological processes at an acceptable cost. Efforts have included engineering organisms to produce precursor products (or even finished products) using the powerful new tools of synthetic biology.[25,26] The Defense Advanced Research Projects Agency (DARPA) and other federal agencies have supported a variety of approaches, which included bio-engineered systems that can produce benzene, toluene, xylenes, polyester, polystyrene, polyurethane, nylon, styrene, butyl rubber, protein concentrates, solvents, resins, plastics, flavoring, lubricants, and many other products.[27] Success is tantalizingly close but impossible to predict.

Biological materials offer an extraordinary range of mechanical properties and environmental durability. As understanding of these materials improves, the staggering complexity of the designs from the nanoscale to macro-scale comes into focus. In nature these structures are manufactured at room temperature from simple materials like calcium carbonate, silica (glass) and calcium phosphate. Finding ways to mimic these systems outside of living organisms will clearly be challenging. Additive manufacturing combined with synthetic biology may eventually provide solutions.[28,29]

Industries that produce and process food and beverages use nearly 3% of EU energy.[30] Efficient water management, pest control and other measures can achieve significant savings but recent innovations may lead to more radical production changes. Synthetic meat production uses innovative manufacturing methods to produce chicken, cheese, beef and other products otherwise produced by animals. Ventures have struggled for years to get products that sufficiently mimic the taste of natural products to be produced at low costs. However the field continues to attract major investments.[31] This market is important since global emissions from livestock production account for more than 14% of all anthropogenic emissions (5% of CO_2, 44% of CH_4, and 53% of N_2O).[32] An industry-sponsored study by the University of Michigan found that replacing conventional beef production with a synthetic process could reduce energy use by 50% and greenhouse gas emissions by tenfold (livestock produce significant amounts of methane).[33]

2.4.4. *Carbon capture*

There has been significant recent interest in finding production systems that actually remove carbon dioxide from the atmosphere. One approach is to capture carbon dioxide generated in large facilities. Cement manufacturing offers opportunities for integrating carbon dioxide capture into conventional production processes, and demonstration projects are underway in Denmark, Norway, Australia and the US.[34]

It may also be possible to make carbon capture an integral part of production chemistry. A recent demonstration captured up to 80% of the carbon dioxide produced in large steelmaking facilities.[35] There are also attempts to achieve synthetic production of hydrocarbons[36] and other materials in processes that take carbon dioxide out of the atmosphere. An X-prize competition, now underway, hopes to accelerate innovation in this field. Finalists in this competition include methods for capturing carbon dioxide to produce "products and materials like concrete, carbon fiber, polymers, food, fertilizer, liquid fuels, graphene, and many others. These have the potential to offer superior performance, lower cost and lower carbon footprint when made from carbon dioxide".[37]

3. The Three Largest Consumers of Industrial Energy

The integrated impact of the process innovations can be illustrated by examining how they can be used in the three largest energy consumers in industry: chemicals, iron and steel, and cement.

3.1. *Chemicals (13% of OECD's carbon dioxide emissions)*[38]

Responsible for nearly 30% of global energy use, the production of chemicals is a major challenge to any effort to cut fossil energy use. Refineries producing gasoline and other fuels use an additional 14% of global energy use. With low cost fossil fuels, plastics and other petrochemicals became the materials of choice for everything from packaging to automobile components. The scale of these operations is enormous. The International Energy Agency (IEA) estimates that petrochemicals will drive a third of the growth in oil demand through 2030.[39] The industry's use of fossil fuels results both from the high temperature processes used in most production and the fact that coal, oil and natural gas are its primary raw materials.

The efficiency of any chemical production process can be increased recovering and re-using waste heat, using cogeneration, and possibly capturing carbon dioxide from large facilities. There are three options for reducing fossil energy use: (1) using biomass or recycled materials as feedstocks (e.g., making fuels from wood waste), (2) using biosynthetic processes described earlier (e.g., engineer bacteria to produce chemical products), and (3) combining zero-carbon hydrogen with carbon reduced from atmospheric carbon dioxide or other sources. Direct use of biomass would have to compete with many other markets for the product in a carbon-constrained economy.

Fossil fuels could continue to power chemical production in a carbon-constrained economy, only if it proves possible to capture and permanently sequester the carbon dioxide and other greenhouse gases they produce. This option is being actively explored and several approaches appear promising.[40] Several of the Carbon X-prize finalists mentioned earlier involve this process. The production of ammonia provides an interesting example of new directions — living plants can capture nitrogen taken directly from air at ambient temperatures and pressures but synthetic ammonia production, principally the Haber–Bosch process, which revolutionized agricultural production in the 20th century, operates at high temperatures and is very energy intensive. Recent innovations have demonstrated that an electricity-based system can produce ammonia from atmospheric nitrogen at ambient temperatures and pressures.[41] The path from these early experiments to the massive scale of commercial petrochemical production will not be easy, but the search for ways to produce chemicals without using fossil fuels remains one of the greatest technological challenges in energy efficiency.

Refineries are traditionally treated differently than chemicals and petrochemicals because their primary product is liquid fuel, which is mainly used for transportation. Refinery energy use and emissions would be largely eliminated if transportation systems shift to carbon-free electricity, biomass, or other sources. These zero-carbon fuel substitutes would also need to operate at an enormous scale.

3.2. *Iron and steel (28% of OECD's carbon dioxide emissions)*

Iron production drove a technological revolution in 500–800 BCE and steel was at the core of the industrial revolution two millennia later. Iron and steel production remain major parts of modern economies and major contributors to global greenhouse emissions. Changing economic structures and market conditions have greatly reduced steel production in developed nations and three quarters of global production is now in China or developing economies.[42]

Steelmaking from raw iron ore, typically beginning in blast furnaces, results in carbon dioxide (CO_2) emissions from the fuel used to heat the material and the fundamental chemistry of ironmaking, namely reducing iron oxide (Fe_2O_3) to pure iron (Fe):

$$Fe_2O_3 + 3CO \rightarrow 2Fe + 3CO_2.$$

There are major opportunities to further improve the efficiency of the process since the theoretical potential efficiency is 350–370 kwh/t-steel, while operating units average 300–550 kwh/t-steel.[43] There is worldwide interest in improving blast furnace efficiency and dramatically reducing emissions.[44] Methods being explored include: improved process control using machine learning, improved sensors and controls, oxyfuel burners, post-combustion optimization, slag preheating, and many others.[45] Tata Steel's Hisarna process claims that carbon dioxide production can be reduced by more than 50% using a system that combines steelmaking process steps into a continuous process that eliminates repeated reheating.[46] There is also interest in cutting carbon dioxide emissions by capturing the gas as it is exhausted from the production process.

Using biomass instead of fossil fuels for power would all but eliminate greenhouse gases. Wood, of course, was used exclusively for steelmaking before the industrial revolution. There will, however, be a large demand for biomass in other sectors in any climate policy that achieves current goals.[47,48] Other savings can result by replacing steel with other materials (see the later discussion) and by using innovative steel alloys that provide much greater performance per unit of steel.[49]

There are, however, steelmaking methods that all but eliminate fossil fuel use and greenhouse gas emissions. Electric arc furnaces, which primarily use scrap, use less than half the energy used in conventional blast furnaces. Net emissions can approach zero if electricity comes from a zero-carbon source. Use of these electricity processes is, however, constrained by the availability of scrap and the fraction (~26%) of steel coming from electric arc processes, which have remained constant for over a decade.[50] Hopefully this can be improved. Technologies that could lower the cost of removing contaminants from scrap would help.

Other methods offer ways to produce steel from ore using methods that could transform or even replace the blast furnace technologies used for most global steel production:[51]

- Capturing and re-using the hot gas escaping during production and to return the captured energy to steel production or to drive another process (co-production of ethanol using biomass waste as a fuel additive).[52,53]
- Direct reduction of ore using hydrogen: $Fe_2O_3 + 3H_2 \rightarrow 2Fe + 3H_2O$,[54] a process that could eliminate emissions depending on the source of the hydrogen.
- Direct iron production with electricity.[55]

3.3. *Cement (27% of OECD's carbon dioxide emissions)*

The rapidly growing cement industry is a major energy consumer responsible for 5–7% of global anthropogenesis carbon dioxide in 2009.[56] Carbon dioxide is released by the energy needed to heat materials to high temperatures and the chemical process that converts the raw material dolomite ($CaMg(CO_3)_2$) to the calcium carbonate ($CaCO_3$) in concrete

$$CaMg(CO_3)_2 \rightarrow CaCO_3 + MgO + CO_2.$$

This carbon dioxide production is partially offset by the fact that cement materials absorb significant amounts of carbon dioxide over a period of decades. One recent estimate suggests that 43% of the carbon dioxide released by cement production between 1930 and 2013 has been reabsorbed.[57] Unlike other industrial sectors, there are no obvious ways to electrify any significant fraction of the production process.

Reduction in energy use and emissions can result from more efficient processes, using a higher fraction of blending with materials (such as fly ash or rice husk ash), new binding materials (calcium hydroxide and silica), capturing waste heat, capturing carbon dioxide for reuse or disposal, and by integrating carbon dioxide capture directly into the production process.[58] There are clear opportunities for improving the productivity of existing processes. The theoretical minimum energy use for producing a ton of cement is about a third of the energy now used in dry kiln processes and about half of what is now used in wet kilns. However efficiency improvements of about 1.5 times the theoretical minimum appear to be practical using known technology.[59,60] Basic research is underway to search for more radical approaches. Cement-like materials like shells and bone can be built by biological processes and efforts are underway to mimic them in synthetic processes.

An obvious way to reduce concrete emissions is to find other materials that can provide equivalent services at an acceptable cost. A number are being explored, but new concrete materials face steep regulatory barriers since failures would lead to major hazards. It is likely that a variety of substitute products will emerge, each tailored to a different market. Concrete materials with low carbon binders, for example, can provide superior insulating value but reduced compressive strength.[61]

4. New Materials

The previous discussion covered ways that new materials such as materials for new separation membranes, thermo-electric generators, and high-temperature coatings can increase the efficiency of industrial production. But efficiencies can also be improved if the manufactured products change — if energy intensive materials can be replaced with materials that require much less energy (or emissions) to produce, or this results in savings when they are in use. Composites, ceramics and bio-inspired materials can replace conventional metals and offer improved characteristics such as resistance to corrosion, performance at high temperatures, improved thermo-electric capabilities, being fracture resistant and self-healing.

The net impact of these materials on energy use, however, depends on trade-offs that can be complex. Savings depend on the energy needed to produce the material, the way new material properties can influence designs, their compatibility with efficient fabrication techniques, the way new products affect the energy efficiency of products that employ the products (e.g., in aircrafts or windows), and the ease with which the materials can be recycled.

Weight savings are of particular interest in transportation equipment. Reducing a car's mass by 10% can, for example, reduce fuel consumption by 6–8%.[62] Carbon fiber and composite materials are used increasingly in aircrafts and other vehicles as replacements for steel and aluminum. The challenge, of course, is that production of these materials relies heavily on petrochemicals, including resins, polyester or vinyl ester.

Biological systems can produce materials with a staggering array of properties such as remarkable strength, toughness and environmental resistance, particularly when compared with other materials of similar weight. Some are self-healing. Others change shape or properties in response to the environment or control signals. It is fascinating to consider, however, what could be done by combining biological design concepts with the materials and processing tools of modern industry. An enormous range of mechanical, electrical and optical properties could be imagined.[63–65]

Wood can provide high strength, earthquake resistance and even superior fire resistance if used in skillful designs. And, of course, wood sequesters carbon dioxide. Wood buildings over 50 meters tall have been built in Norway and British Columbia, and Vienna has a new building that is 84 meters tall. The Sakyamuni Pagoda in China, built in 1056, is 67 m tall and is still standing.[66]

5. Recycling

Use of recycled materials can greatly reduce the energy used in production. For some metals, the amount of metals in use or in scrap approaches the amount still unmined, raising the possibility that recycled metals might be able to meet a significant fraction of all future supplies.[67]

In most cases, the barriers to increased recycling are economic or political and not technical. Low energy, raw material costs and the absence of policy incentives in many areas present clear problems. Recycling rates vary enormously by region.

Recycling is less expensive when products are designed to make disassembly easier. In the case of plastics, greater use of recyclable materials and greater uniformity of products would also increase useful recycling rates.

Recycling		
	Energy efficiency gain (% savings from pro production from raw materials[a])	Recycling rate
Iron and steel	60–70%	33%[b]
Aluminum	90%	18.5%
Plastics	—	9% (US),[c] 30% (EU)[d]

Notes: [a]IEA, 2018. [b]US EPA, "Advancing sustainable materials management: 2015 fact sheet", United States Environmental Protection Agency, 2018. https://www.epa.gov/sites/production/files/2018-07/documents/2015_smm_msw_factsheet_07242018_fnl_508_002.pdf. Accessed: 14 January 2019. [c]A.H. Tullo, "Should plastics be a source of energy?" *C&EN*, 2018. https://cen.acs.org/environment/sustainability/Should-plastics-source-energy/96/i38. Accessed: 14 February 2019. [d]EU, "A European strategy for plastics in a circular economy", The European Commission, Brussels, 2018. http://ec.europa.eu/environment/circular-economy/pdf/plastics-strategy.pdf. Accessed: 14 February 2019.

Increased use of recycled products relies on policies that would encourage the production of recyclable products and the collection of the products. Thermoplastics, for example, can be re-melted and reused while thermoset plastics cannot be recycled. Recycling is ultimately limited by the fact that the value of the material declines rapidly after repeated cycles.[68] At present, plastics can only be recycled a few (<10) times. Invention of materials that can either be successfully reused or biodegraded in all environments would be a major step forward (many "biodegradable" materials now in use do not degrade in oceans).[69] Benefits would be even greater if these materials could be made from net negative carbon dioxide processes.

6. Conclusion

The world economy is moving rapidly away from energy and material intensity, with services and high-value/low weight products dominating economic output. But the energy needed to produce products remains a large fraction of global energy use and greenhouse gas production. Meeting global climate goals means, however, that energy efficiency must increase much faster than demand increases. In the case of industry this means innovations covering a huge range of products and processes. The previous discussion has shown that there is a range of innovations that could meet this challenge. An aggressive global commitment to funding research, and using carbon pricing and other mechanisms to create a market for inventions will be essential for ensuring that needed technologies enter the market in time to meet pressing environmental goals.

References

1. US DOE, "International energy outlook 2016", US Department of Energy, EIA, 2016. https://www.eia.gov/outlooks/ieo/pdf/industrial.pdf. Accessed: 16 February 2019.
2. IEA, "Energy efficiency 2018", International Energy Agency, 2018. https://webstore.iea.org/download/direct/2369?fileName_Market_Report_Series_Energy_Efficiency_2018.pdf. Accessed: 7 February 2019.
3. IPCC, "Global Warming of 1.5°C", Intergovernmental Panel on Climate Change, 2018. https://report.ipcc.ch/sr15/pdf/sr15_spm_final.pdf.
4. M. Rehfeldt, T. Fleiter and F. Toro, "A bottom-up estimation of the heating and cooling demand in European industry", *Energy Efficiency* **11** (2018) 1057. https://www.researchgate.net/publication/312173884_A_bottom-up_estimation_of_heating_and_cooling_demand_in_the_European_industry.
5. President's Council of Advisors on Science and Technology, "Report to the President: accelerating US advanced manufacturing, Executive Office of the President", Washington, 2014. https://www.manufacturingusa.com/sites/prod/files/amp20_report_final.pdf. Accessed: 7 February 2019.
6. S. Nimbalkar *et al.*, "Smart manufacturing technologies and data analytics for improving energy efficiency in industrial energy systems", *2017 ACEEE Summer Study on Energy Efficiency in Industry*, Denver, 15 to 18 August 2017.
7. US DoE, "Quadrennial technology review", US Department of Energy, 2015. https://www.energy.gov/sites/prod/files/2017/03/f34/qtr-2015-chapter6.pdf. Accessed: 7 February 2019.
8. *Ibid.*

9. C. Burt, X. Piao, F. Gaudi, B. Busch and N.F.N. Taufik, "Electric motor efficiency", ITRE report, 2006. http://www.itrc.org/reports/pdf/r06004.pdf. Accessed: 14 February 2019.

10. M. Lowe, R. Golini and G. Gereffi, "US adoption of high-efficiency motors and drives: lessons learned", 2010 Duke University Center on Globalization, Governance & Competitiveness.

11. USA and the International Institute for Applied Systems Analysis, *Global Energy Assessment — Toward a Sustainable Future* (Cambridge University Press, Cambridge). http://www.iiasa.ac.at/web/home/research/Flagship-Projects/ Global-Energy-Assessment/Chapters_Home.en.html. Accessed: 16 February 2019.

12. International Energy Agency, "Wide Band Gap Technology: Efficiency Potential and Application Readiness Map", 2020, International Energy Agency. https://www.iea-4e.org/document/447/wide-band-gap-technology-efficiency-potential-and-application-readiness-map.

13. US Department of Energy, Quadrennial Technology Review, 2015. https:// www.energy.gov/sites/prod/files/2017/03/f34/qtr-2015-chapter6.pdf.

14. *Ibid.*

15. T. Hettesheimer, S. Hirzel and H.B. Roß, "Energy savings through additive manufacturing: an analysis of selective laser sintering for automotive and aircraft components", *Energy Efficiency* **11** (2018) 1227. https://doi.org/ 10.1007/s12053-018-9620-1.

16. US Department of Energy, Quadrennial Technology Review, 2015. https:// www.energy.gov/sites/prod/files/2017/03/f34/qtr-2015-chapter6.pdf.

17. M. Rehfeldt, T. Fleiter and F. Toro, *Energy Efficiency* **11** (2018) 1057. https:// www.researchgate.net/publication/312173884_A_bottom-up_estimation_ of_heating_and_cooling_demand_in_the_European_industry.

18. US DOE, "Split-system cold climate heat pump", US Department of Energy. https://www.energy.gov/eere/buildings/downloads/split-system-cold-climate-heat-pump. Accessed: 15 February 2019.

19. G. Schumm *et al.*, "Hybrid heating system for increased energy efficiency and flexible control of low temperature heat", *Energy Efficiency* **11** (2018) 1117. https://doi.org/10.1007/s12053-017-9584-6.

20. J. Jutsen, A. Peers and L. Hutton, "High temperature heat pumps for the Australian food industry", Australian Alliance for Energy Productivity (A2EP), 2017. https://www.airah.org.au/Content_Files/Industryresearch/ 19-09-17_A2EP_HT_Heat_pump_report.pdf.

21. Denis-Ryan, C. Bataille and F. Jotzo, "Managing carbon-intensive materials in a decarbonizing world without a global price on carbon", *Climate Policy*, **16** (2016). https://doi.org/10.1080/14693062.2016.1176008.

22. The National Academies, *Real Prospects for Energy Efficiency in the United States* (National Academies Press, Washington, 2010). https://www.nap.edu/ read/12621/chapter/6. Accessed: 7 February 2019.

23. J.A. Moses, T.J. Norton, K. Alagusundaram and T.B. Kumar, "Novel drying techniques for the food industry", *Food Engineering Reviews* **6** (2014) 43–55. DOI 10.1007/s12393-014-9078-7.

24. I.S. Al Rowaihi *et al.*, "A two-stage biological gas to liquid transfer process to convert carbon dioxide into bioplastic", *Bioresource Technology Reports* **1** (2018) 61–68.

25. R.A. Le Feuvre and N.S. Scrutton, "A living foundry for synthetic biological materials: A synthetic biology roadmap to new advanced materials", *Synthetic and Systems Biotechnology* **3** (2018) 105–112. DOI: 10.1016/j.synbio.2018.04.002.

26. BIO, "Current uses of synthetic biology", Biotechnology Innovation Organization, 2019. https://www.bio.org/articles/current-uses-synthetic-biology.

27. B. Wolfson, "DARPA and the future of synthetic biology", *O'Reilly*, 2017. https://www.oreilly.com/ideas/darpa-and-the-future-of-synthetic-biology. Accessed: 8 February 2019.

28. R.A. Le Feuvre and N.S. Scrutton, "A living foundry for synthetic biological materials: A synthetic biology roadmap to new advanced materials, *Synthetic and Systems Biotechnology* **3** (2018) 105–112. DOI: 10.1016/j.synbio.2018.04.002.

29. U.G.K. Wegst *et al.*, "Bioinspired structural materials", *Nature Materials* **14** (2015) 23–36. DOI: 10.1038/nmat4089.

30. Eurostat, "Agri-environmental indicator-energy use", https://ec.europa.eu/eurostat/statistics-explained/index.php/Agri-environmental_indicator_-_energy_use. Accessed: 15 February 2019.

31. G.W. Schaefer, "Lab-grown meat", *Scientific American*, 2018 https://www.scientificamerican.com/article/lab-grown-meat/. Accessed: 11 February 2019.

32. P.J. Gerber *et al.*, "Tackling climate change through livestock — a global assessment of emissions and mitigation opportunities", Food and Agriculture Organization of the United Nations, 2013. http://www.fao.org/3/a-i3437e.pdf. Accessed: 11 August 2020.

33. M.C. Heller and G.A. Keoleian, "Beyond Meat's beyond burger life cycle assessment", University of Michigan Center for Sustainable Systems, 2018. http://css.umich.edu/publication/beyond-meats-beyond-burger-life-cycle-assessment-detailed-comparison-between-plant-based. Accessed: 11 February 2019.

34. IEA, "Technology roadmap, low-carbon transition in the cement industry", International Energy Agency. https://webstore.iea.org/technology-roadmap-low-carbon-transition-in-the-cement-industry. Accessed: 12 February 2019.

35. IEA, "Iron and steel", International Energy Agency, 2020 https://www.iea.org/reports/iron-and-steel. Accessed: 11 August 2020.

36. S.J. Davis, "Net-zero emissions energy systems", *Science*, 2018. DOI: 10.1126/science.aas9793.

37. NRG COSIA Carbon XPRIZE, https://assets-us-01.kc-usercontent.com/5cb25086-82d2-4c89-94f0-8450813a0fd3/ec5aba69-e68b-48c8-99b0-151e21749d67/

XPRIZE%20Carbon%20Finalist%20Team%20Deck.pdf. Accessed: 8 February 2019.

38. IEA, "Tracking 2017", International Energy Agency. https://webstore.iea.org/tracking-clean-energy-progress-2017. Accessed: 11 August 2020.

39. IEA, "The future of petrochemicals", International Energy Agency, 2018, https://webstore.iea.org/download/direct/2310?fileName=The_Future_of_Petrochemicals.pdf. Accessed: 12 February 2019.

40. F.S. Zeman and D.W. Keith, "Carbon neutral hydrocarbons", *Philosophical Transactions of the Royal Society A*, (2008). https://doi.org/10.1098/rsta.2008.0143.

41. R. Lan, J.T.S. Irvine and T. Shanwen, "Synthesis of ammonia directly from air and water at ambient temperature and pressure", *Nature Scientific Reports* **3** (2013). DOI: 10.1038/srep01145.

42. Worldsteel Association, "Steel's contribution to a low carbon future and climate resilient societies", *Nuclear Energy Efficiency* **11** (2018) 1083.

43. Institute for Industrial Productivity, "Industrial efficiency technology database", Institute for Industrial Productivity. http://www.iipinetwork.org/resources/industrial-efficiency-technology-database. Accessed: 11 August 2020.

44. J. Yan, "Progress and the future of breakthrough low-carbon steelmaking technology (ULCOS) of EU", *International Journal of Mineral Processing and Extractive Metallurgy,* (2018) 15–22. DOI: 10.11648/j.ijmpem.20180302.11.

45. http://www.iipinetwork.org/wp-content/Ietd/content/electric-arc-furnace.html.

46. Tata Steel, "Hisarna: game changer in the steel industry". https://www.tatasteeleurope.com/static_files/Downloads/Corporate/About%20us/hisarna%20factsheet.pdf. Accessed: 12 February 2019.

47. IPCC, 2018.

48. IEA, Energy Technology Perspectives, Institute for Industrial Productivity, 2017.

49. C.M. Sonsino, "Light-weight design chances using high-strength steels", Fraunhofer Institute for Structural Durability and System Reliability. https://www.phase-trans.msm.cam.ac.uk/2005/LINK/171.pdf. Accessed: 13 February 2019.

50. IEA, 2019.

51. *Ibid.*

52. *Ibid.*

53. *Ibid.*

54. V. Vogl, M. Ahman and L.J. Nilsson, "Assessment of hydrogen direct reduction for fossil-free steelmaking", *Journal of Cleaner Production,* (2018), 736–745. https://doi.org/10.1016/j.jclepro.2018.08.279.

55. J. Wiencke, "Electrolysis of iron in a molten oxide electrolyte", *Journal of Applied Electrochemistry* **48** (2018) 115–126. https://link.springer.com/article/10.1007/s10800-017-1143-5.

56. R. Maddalena, J.J. Roberts and A. Hamilton, "Can Portland cement be replaced by low-carbon alternative materials?" *Journal of Cleaner Production*, (2018) 933–942. https://doi.org/10.1016/j.jclepro.2018.02.138.

57. F. Xi *et al.*, "Substantial global carbon uptake by cement carbonation", *Nature Geosciences* **9** (2016) 880–883. http://doi.org/10.1038/ngeo2840.

58. IEA, "Low carbon transition in the cement industry", International Energy Agency. 2018, https://www.iea.org/reports/technology-roadmap-low-carbon-transition-in-the-cement-industry.

59. *Ibid.*

60. Natural Resources Canada, "Energy consumption benchmark guide: cement", 2001. https://www.nrcan.gc.ca/energy/publications/efficiency/industrial/6005. Accessed: 12 February 2019.

61. Maddalena, R, Roberts, J.J., Hamilton, A, Can Portland cement be replaced by low-carbon alternative materials? A study on the thermal properties and carbon emissions of innovative cements, Journal of Cleaner Production Volume 186, 10 June 2018, Pages 933–942, https://doi.org/10.1016/j.jclepro.2018.02.138.

62. US Department of Energy, Quadrennial Technology Review, 2015. https://www.energy.gov/sites/prod/files/2017/03/f34/qtr-2015-chapter6.pdf.

63. R.A. Le Feuvre and N.S. Scrutton, "A living foundry for synthetic biological materials: A synthetic biology roadmap to new advanced materials", Synthetic and Systems Biotechnology **3** (2018) 105–112. DOI: 10.1016/j.synbio.2018.04.002.

64. Ulrike G. K. Wegst *et al.*, Bioinspired structural materials, *Nature Materials*; London **14** (1) (2015), 23–36. DOI:10.1038/nmat4089.

65. N. Sakhavand and R. Shahsavari, "Universal composition-structure-property maps for natural and biomimetic platelet-matrix composites and stacked heterostructures", *Nature Communications*, **6** (2016). http://doi.org/10.1038/ncomms7523.

66. J. Tollefson, "The wooden skyscrapers that could help to cool the planet", *Nature*, Vol. 545, 2017. https://www.nature.com/news/the-wooden-skyscrapers-that-could-help-to-cool-the-planet-1.21992#correction1. Accessed: 8 February 2019.

67. R.B. Gordon, M. Bertram and T.E. Graedel, "Metal stocks and sustainability", *Proceedings of the National Academy of Sciences of the United States of America*, (2006) 1209–1214. https://doi.org/10.1073/pnas.0509498103.

68. EU, "A European strategy for plastics in a circular economy", The European Commission, Brussels, 2018. http://ec.europa.eu/environment/circular-economy/pdf/plastics-strategy.pdf. Accessed: 14 February 2019.

69. BIO, 2019.

Chapter 4

Next Generation Mobility Systems

Susan Shaheen[*,‡] and Adam Cohen[†,§]

*Department of Civil and Environmental Engineering at the
University of California, Berkeley

†Transportation Sustainability Research Center University of California,
Berkeley, California 94720, USA

‡sshaheen@berkeley.edu

§apcohen@berkeley.edu

1. Introduction

In recent years, mobility on demand (MOD) is gaining popularity among mobility consumers. This innovative concept is based on the principle that transportation is a commodity where modes have economic values that are distinguishable in terms of cost, journey time, wait time, number of connections, convenience, and other attributes. MOD enables consumers to access mobility, goods and services on demand by dispatching or using shared mobility, delivery services and public transportation strategies through an integrated and connected multi-modal network. This chapter describes the different services that have emerged in the MOD ecosystem and the core enablers of MOD, such as stakeholders, business models and technology. The chapter concludes with a discussion of how MOD and vehicle automation could impact cities and the transportation network and their energy use.

In recent years, technological, economic and environmental forces are changing the way people travel and consume resources.[1,2] On-demand mobility and delivery services have become part of a sociodemographic trend that has pushed MOD from the fringe toward the mainstream.[3] MOD is changing how people travel and access goods, and it is having a

155

transformative effect on cities.[4] This concept has also been referred to as mobility as a service (MaaS). This chapter has eight sections. The first two sections define MOD and discuss the similarities and differences between MOD and MaaS. Section 2 reviews the MOD ecosystem including the role of supply and demand and MOD stakeholders. In Section 3, a taxonomy of shared mobility and goods delivery is reviewed. In Section 5, business models and technology enablers are discussed. Section 5 concludes with a discussion of how MOD and vehicle automation could impact cities and the transportation network. Section 6 summarizes key findings of this chapter.

2. The Mobility on Demand Ecosystem

MOD is an innovative concept based on the principle that transportation is a commodity where modes have economic values that are distinguishable in terms of cost, journey time, wait time, number of connections, convenience, and other attributes.[3,5] MOD enables consumers to access mobility, goods and services on demand by dispatching or using shared mobility, delivery services and public transportation strategies through an integrated and connected multi-modal network (see Figure 1).[3] The most advanced forms of MOD passenger services incorporate trip planning and booking, real-time information and fare payment into a single user interface. Passenger modes facilitated through MOD include: bikesharing, carsharing, ridesharing (carpooling and vanpooling), transportation network companies (TNCs) (also known as ridesourcing and ridehailing), scooter sharing (including standing electric and moped-styles), microtransit, shuttle services, public transportation and other emerging transportation strategies. The most advanced forms of MOD incorporate robotic delivery, app-based courier network services (CNS), and aerial passenger and delivery services (e.g., advanced air mobility, such as urban air mobility and drone delivery). Fundamentally, MOD is about how people make mobility decisions, how they move, how they consume goods and services, and the stakeholders that make it possible.[3]

Commodification is the transformation of transportation services into tradable commodities or an economic resource.[3,6] As part of this transformation, transportation services are assigned an economic value that can be traded or exchanged in the transportation marketplace for other transportation mode(s).[5,7] MOD emphasizes enhancing mobility options for all users through the integration of on-demand modal services, public transportation, payment mechanisms, traveler incentives and an array of

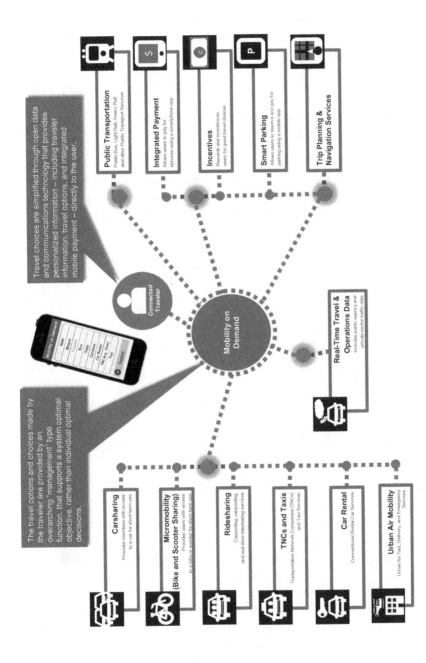

Fig. 1. User-centric travel options.

Source: Adapted from US Department of Transportation.

real-time information services.[3] MOD is focused on providing travelers with more seamless travel options (i.e., routing, booking and payment) for all trip segments. This seamless integration improves the user experience and can enable more informed and sustainable transportation choices.

3. What is Mobility as a Service and How Does it Differ from MOD?

MOD differs from another concept, MaaS. MOD examines the commodification of passenger mobility and goods delivery and transportation systems management, whereas MaaS primarily focuses on passenger mobility aggregation and subscription services (see Figure 2).[8] Specifically, MaaS is a transportation concept that integrates existing and innovative mobility services into one single digital platform where customers purchase mobility service packages tailored to their individual needs (ranging from per trip fares to bundled subscription mobility services).[9] Brokering travel with suppliers, repackaging and reselling it as a bundled package is a distinguishing characteristic of MaaS.

Utriainen and Pöllänen[10] conducted a literature review of MaaS by identifying 37 peer-reviewed journal articles and conference papers. Key characteristics of MaaS identified across this literature include:

1. The integration of traditional and innovative transportation modes (i.e., shared mobility)[11,12];
2. The option for pay-as-you-go and subscription pricing[13];
3. A single platform where users can plan, book, pay and get tickets for their trips[9,14–16];
4. Multiple stakeholders engage in MaaS (customers, service providers, apps, public agencies, etc.)[16–19];
5. The use of information and communications technologies (e.g., smartphone apps)[19–21]; and
6. A customized mobility experience that allows users to modify available trips based on traveler preferences.[15]

Sochor and colleagues[22] established a MaaS typology describing four levels of varying integration: level 0 — no integration; level 1 — information integration; level 2 — booking and payment integration; level 3 — service offer integration; and level 4 — the integration of societal goals. Sochor

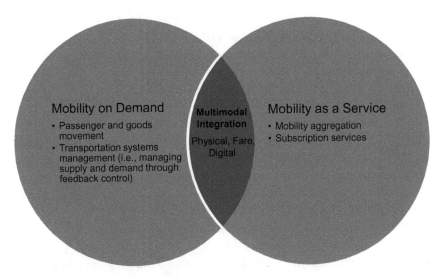

Fig. 2. Similarities and differences of MOD and MaaS.

Source: S. Shaheen and A. Cohen, "Mobility on Demand (MOD) and Mobility as a Service (MaaS). How are they similar and different?" *Move Forward*, 7 March 2019. https://www.move-forward.com/mobility-on-demand-mod-and-mobility-as-a-service-maas-how-are-they-similar-and-different/.

describes level 1 as primarily travel planning funded through advertising or taxpayer funds. Level 1 service providers aggregate and display data but do not have a fiduciary responsibility to ensure data fidelity. With level 2, service providers integrate trip booking and payment to enhance customer convenience and encourage multimodal travel. For service providers, level 2 grows the potential customer base but also increases potential competition by offering transportation services alongside other service providers. Because level 2 integrates ticketing and payment, data fidelity is key. Level 3 is intended to serve as a comprehensive alternative to private vehicle ownership by bundling transportation services together and offering subscription packages. According to Sochor and colleagues,[22] level 3 emphasizes meeting a household's complete mobility needs rather than a single trip between an origin and destination. Level 4 adds value by employing incentives, gamification and other policies to impact traveler choices and influence societal and environmental outcomes. Sochor and colleagues[22] conclude that lumping all MaaS services under a single loosely defined concept can create confusion and possibly undermine the intended concept, thereby necessitating a framework to blend technological and institutional integration of MaaS services.

In summary, the literature generally concludes that MaaS is a concept about integrating multiple transportation modes into a seamless user experience, requiring open data and cooperation by public and private transportation stakeholders.

MaaS is primarily focused on mobility aggregation and subscription services for passenger mobility options where MOD leverages passenger mobility and goods delivery services to enhance access while simultaneously leveraging MOD to achieve transportation system operational improvements by helping agencies balance supply and demand to match changing conditions across the transportation network.[23]

With MOD, public agencies have the potential to:

- Embrace the needs of all users (travelers and shippers), public and private facilities, and services across all modes — including motor vehicles, pedestrians, bicycles, public transit, for-hire vehicle services, carpooling/vanpooling, goods delivery and other transportation services[3];
- Improve the efficiency of the transportation system (including energy efficiency) and increase the accessibility and mobility of all travelers[3];
- Enable transportation system operators and their partners to monitor, predict and influence conditions across an entire mobility ecosystem and for an entire region[3]; and
- Receive data inputs from multiple sources and provide response strategies geared to various operational objectives.[3]

MOD has three major guiding principles: (1) traveler centric and consumer driven, (2) data connected and platform independent, and (3) multimodal and mode agnostic.[3] MOD promotes choice in personal mobility, leverages emerging and existing technologies and big data capabilities, encourages multimodal connectivity and system interoperability, and promotes new business models that enhance traveler experience.

4. The MOD Ecosystem

Figure 3 demonstrates how MOD can interact and influence supply and demand of the transportation network. Leveraging big data, multimodal transportation operations receives input from the rest of the system and influences it through feedback control mechanisms that help manage supply and demand.[3,4] The components of the supply and demand side are based on

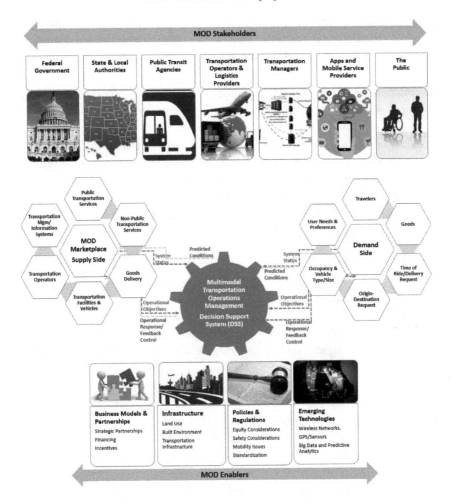

Fig. 3. Supply and demand sides of MOD ecosystem.

Source: S. Shaheen, A. Cohen, B. Yelchuru and S. Sarkhili, *Mobility on Demand Operational Concept Report* (US Department of Transportation, Washington, 2017).

consumption choice and trip generation. The supply side of this ecosystem consists of all the players, operators and devices that provide transportation services for people or goods delivery including:

- Public transportation services;
- Non-public transportation services including: shared micromobility (bikesharing and scooter sharing), carpooling, car rentals, carsharing, microtransit, TNCs, taxis and other private mobility suppliers;

- Goods delivery services including freight, logistics, first-and-last mile goods delivery, courier network services, drones and robotic delivery;
- Transportation facilities including parking, tolls, roadways and highways;
- Vehicles of all types such as public transit vehicles, private vehicles, goods delivery vehicles and emergency vehicles that could be connected and automated;
- Transportation management and information systems such as payment systems for parking, tolling and public transit, signal systems, mobile apps for trip planning and payment (for all travelers), fleet management systems and navigation systems; and
- Public and private transportation information services including schedule information, 511 and dynamic message signs, Google maps, etc.

The demand side of this ecosystem consists of all the system users (travelers and couriers) and their choices and preferences including:

- Pedestrians, riders, drivers, cyclists (reflecting the wide spectrum of demographic users, including older adults, people with disabilities, children, etc.);
- Goods and merchandise requiring physical delivery;
- Time of ride and/or delivery request;
- Origin-destination request, which determines the location of the demand and affects the route and mode choice;
- Modal demand based on occupancy, size or type of vehicle requested; and
- User needs and preferences including mode and decision choices on how a trip is made (such as decisions to drive alone, carpool, use public transport, or some other form of shared mobility options).

There is a wide range of modal stakeholders that enable MOD.[3] Common MOD stakeholders include:

- Federal government agencies that play a role in establishing transportation strategies, policies and legislation;
- State and local authorities including municipalities and metropolitan planning organizations (MPOs) that play a role in implementing policy

and regulations, issuing permits, managing public transport in the region and improving transportation operations. They also provide strategic urban planning and traffic planning and are responsible for the local infrastructure;

- Public transit agencies, such as city buses, trolley buses, trams (or light train), rapid transit (metro, subway), ferries and paratransit;
- Transportation and traffic managers, including transportation management centers that monitor operations, allocate resources as necessary and respond to network needs;
- Transportation and logistics service providers that offer mobility and/or delivery services;
- Apps and mobile service providers that enable on-demand mobility and delivery (e.g., mobile ticketing, payment, navigation services, etc.); and
- Consumers.

5. Mobility and Delivery Services

MOD stakeholders can have a variety of similar and differing roles, such as: (1) commoditizing passenger mobility and goods delivery; (2) offering short-term, on-demand access to mobility and goods delivery services for users; (3) enhancing convenience by facilitating trip planning or delivery, payment and other functions into a single interface; (4) providing convenience through additional on-demand mobility and delivery options; (5) providing transportation services for people including individuals with special needs; and (6) increasing mobility and goods availability (e.g., journeys previously inaccessible by a single mode, first-and-last mile connections, additional service offerings during off-peak or high-congestion travel times and access to goods/services previously unavailable).[3,23] In particular, shared mobility service providers and goods delivery services are enablers of the commodified marketplace because they create a larger user pool and modal options that support a "network effect" where modal options are in closer proximity to one another (physical and digital), adding collective value.[7] Together, MOD can unlock a "multimodal multiplier" effect where intermodal synergies are greater together than the sum of the parts. The next two sections explore the shared mobility and goods delivery marketplaces, which represent the core enablers of the MOD supply side.

5.1. *Shared mobility*

Shared mobility — the shared use of a vehicle, bicycle or other travel mode — is an innovative transportation strategy that enables users to have short-term access to a transportation mode on an as-needed basis. Shared mobility includes a number of transportation modes and service models to meet diverse traveler needs. Shared mobility can include roundtrip services (a vehicle, bicycle or other mode is returned to its origin); one-way station-based services (vehicle, bicycle or mode is returned to a different designated station location); and one-way free-floating services (vehicle, bicycle or mode can be returned anywhere within a geographic area).[1,24] Shared mobility modes comprising the MOD ecosystem include:

1. **Bikesharing:** Provides users with on-demand access to bicycles at a variety of pick-up and drop-off locations for one-way (point-to-point) or roundtrip travel. Bikesharing fleets are commonly deployed in a network within a metropolitan region, city, neighborhood, employment center and/or university campus.[25,26] Bikesharing typically includes one of three common service models:

    a. Station-based bikesharing systems where users access bicycles via unattended stations offering one-way station-based service (i.e., bicycles can be returned to any station).

    b. Dockless bikesharing systems where users check out a bicycle and return it to any location within a predefined geographic region. Dockless bikesharing can include business-to-consumer or peer-to-peer systems enabled through third-party hardware and applications.

    c. Hybrid bikesharing systems where users check out a bicycle from a station and end their trip either returning it to a station or a non-station location, or users pick up a dockless bicycle and either return it to a station or any non-station location.

2. **Carsharing:** Individuals gain the benefits of private vehicle use without the costs and responsibilities of ownership. Individuals typically access vehicles by joining an organization that maintains a fleet of cars and light trucks deployed in lots located within neighborhoods and at public transit stations, employment centers, and colleges and universities. Typically, the carsharing operator provides gasoline, parking and maintenance. Generally, participants pay a fee each time they use a vehicle.[27]

3. **Microtransit:** Privately or publicly operated, technology-enabled transit services that typically employ multi-passenger/pooled shuttles or vans to provide on-demand or fixed-schedule services with either dynamic or fixed routing.[24]

4. **Personal Vehicle Sharing:** Sharing of privately owned vehicles where companies broker transactions among car owners and renters by providing the organizational resources needed to make the exchange possible (e.g., online platform, customer support, driver and motor vehicle safety certification, auto insurance, technology). This is often referred to as peer-to-peer carsharing.

5. **Transportation Network Companies (TNCs):** TNCs (also known as ridesourcing and ridehailing) provide prearranged and on-demand transportation services for compensation, which connect drivers of personal vehicles with passengers. Smartphone applications are used for booking, ratings (for both drivers and passengers) and electronic payment.[1,24]

6. **Ridesharing:** Ridesharing (also known as carpooling and vanpooling) facilitates formal or informal shared rides between drivers and passengers with similar origin-destination pairings.[1,24]

7. **Scooter Sharing:** Users gain the benefits of a private scooter without the costs and responsibilities of ownership. Individuals typically access scooters by accessing a fleet of scooters at various locations. Scooter operators typically provide power or gas (as appropriate), maintenance and may provide parking. Participants pay a fee each time they use a scooter.[1,24,26] Trips can be roundtrip or one-way. Scooter sharing includes two types of services:

    a. Standing electric scooter sharing using shared scooters with a standing design including a handlebar, deck and wheels propelled by an electric motor. At present, the most common scooters are made of aluminum, titanium and steel.

    b. Moped-style scooter sharing using shared scooters with a seated-design, is electric or gas powered, and generally has a less stringent licensing requirement than motorcycles designed to travel on public roads.

8. **Taxis:** Taxi services provide prearranged and on-demand vehicle services for compensation through a negotiated price, zone pricing or a taximeter. Trips can be made by advance reservations (booked through a phone, website or smartphone application), street hail (by

raising a hand or standing at a taxi stand or specified loading zone), or e-Hail (dispatching a driver using a smartphone application).[1,24]

9. **Urban Air Mobility:** Urban air mobility (UAM) includes mobility, goods delivery and emergency services, and Unmanned Aerial Systems (UAS) support using a variety of aircraft designs and a mix of onboard, ground-piloted and autonomous operations.[28,29]

5.2. *Urban goods delivery*

Urban goods delivery is equally important to MOD as changing consumer patterns disrupt traveler behavior. In recent years, innovative technologies and business models are helping to reimagine service delivery. Whether it is a startup (e.g., Uber Eats, Postmates, DoorDash), an internet-based retailer (e.g., Amazon) or a supply chain and logistics firm, advancements in courier services (technologies and service models) are transforming consumer behavior and disrupting supply and trip chains (e.g., linking a series of destinations in one single-origin based trip). Travelers' decisions to change their consumption preferences from driving to the store on the way home from work to having goods delivered could drive fundamental changes in travel behavior.[3]

On-demand courier services have grown rapidly due to technology advancements, changing consumer patterns and a growing consumer recognition that goods delivery can serve as substitutes for personal trips to access goods and services. Together these trends have transformed the retail sector from "just in time" inventory, where retailers order inventory and stock shelves on an as-needed basis to "just in time delivery", with goods delivered directly to consumers on-demand.[30] In recent years, subscription e-commerce has grown exponentially led by many models such as: (1) Dollar Shave Club and Ipsy cosmetics, (2) Stitch Fix fashion on-demand, (3) meal kit delivery services including Blue Apron, Hello Fresh, and Home Chef, and (5) grocery delivery services such as AmazonFresh, Postmates and Instacart. This also includes shipping subscription services, such as ShopRunner and Amazon Prime, that offer unlimited priority delivery services for a flat monthly or annual fee. Even companies, such as Wayfair, are offering free shipping on most merchandise, even larger items (e.g., furniture).[3,30] Advanced algorithms are often employed to aid merchants and delivery providers to optimize the supply and delivery chain, ranging from order fulfillment to identifying the least expensive or quickest delivery route. Six innovations in goods delivery that are impacting the MOD ecosystem include:

1. **Subscription Delivery Services:** The growth of low-cost, flat-rate delivery subscriptions (e.g., Amazon Prime and Shop Runner) that allow consumers access to on-demand delivery consumption — a key factor contributing to induced demand.

2. **Locker Delivery:** Locker delivery, which is already widely deployed by the US Postal Service, allows consumers to order and ship items to a self-service locker at home, work or an alternative pick-up location. Locker delivery can help consumers, merchants and delivery providers overcome a variety of challenges, such as weekend and off-peak delivery services and enhanced security (versus leaving a package at a door).

3. **Courier Network Services (CNS):** Apps or online platforms employed to provide for-hire delivery services for monetary compensation. The apps match couriers — who use a personal vehicle, bicycle, or scooter for deliveries — with customers of the ordered goods (i.e., packages, food).

4. **Drones:** A delivery drone is a short-range unmanned aerial vehicle (UAV) that can transport small packages, food or other goods. Some service providers, such as the United Parcel Service, have experimented with pairing drones and truck-based delivery to improve service delivery.

5. **Robotic Delivery:** Like drones, delivery robots offer short-range unmanned ground-based delivery of packages, food or other goods.

6. **Automated Vehicles (AVs):** Automated and connected vehicles offer another mechanism for future delivery options employing both business-to-consumer and peer-to-peer delivery services. Increasingly, last-mile delivery is being reimagined through automated processes. In Summer 2018, Kroger began testing driverless grocery deliveries in Arizona using AVs. Numerous companies, such as Dispatch and Starship, are delivering food, beverages, parcels and other items using small delivery robots. To accept delivery when a person is at work or away, numerous companies are developing innovations to facilitate secure parcel delivery methods that could be used in conjunction with human couriers, AVs or both. A few of these innovations include:

   a. ParcelHome, an electric lockbox that can securely send and receive parcels with any courier service;
   b. Pharme, which enables the delivery of packages to the trunk of a car; and
   c. Amazon Key that enables vehicle trunk and in-home deliveries.

Together the growth of e-commerce, subscription services and last-mile delivery may contribute to a dramatic increase in goods-related trips across the entire transportation network.[30] However, it is not just the growth of e-commerce, subscription services and last-mile delivery that have the potential to increase delivery trips. In an automated future, there could be an increasing number of marketplace players including startups, courier services and retailers (who may more readily opt to operate their own delivery fleets). Goods delivery innovations within the MOD ecosystem have the potential to disrupt both businesses (e.g., retail) and daily travel behavior (e.g., induced demand, congestion).[3,30]

6. MOD Business Models

Fundamentally, these business models can be categorized into four groups based on the MOD service provider and consumer: (1) business to consumer (B2C), (2) business to government (B2G), (3) business to business (B2B), and (4) peer-to-peer (P2P).[1,3] There can be overlap among business models due to variations in the services provided, ownership, administration and operational characteristics.

B2C services provide individual consumers with access to business owned and operated transportation services, such as a fleet of vehicles, bicycles, scooters or other modes through memberships, subscriptions, user fees or a combination of pricing models. B2G offers transportation services to a public agency. Pricing may include a fee-for-service contract, per-transaction basis or some other pricing model. B2B operators sell business customers access to transportation services either through a fee-for-service or usage fees. The service is typically offered to employees to complete work-related trips. P2P services broker mobility and courier transactions among car, bicycle or other device owners and renters by providing the organizational resources needed to make the exchange possible (i.e., online platform, customer support, driver and motor vehicle safety certification, auto insurance and technology).[1,3]

7. Technology as an Enabler: The Role of Data and Smartphone Apps in MOD

Digitization of the transportation network — from real-time analytics, mobile applications, sensors and satellite navigation — allows travelers to be more informed, agile and mobile in their transportation decisions.

Leveraging data and real-time analytics at all stages of the traveler process is a key MOD enabler. Data understanding can aid public agencies and transportation operators (public and private) build a more intelligent, responsive and agile transportation network.[2,3]

The dramatic increase in intelligent transportation systems, location-based wireless and cloud technologies, coupled with the growth of data, are driving fundamental changes in consumer behavior. End users are employing mobile websites and apps for an array of transportation functions, such as vehicle routing and parking, information services, trip planning, fare payment and goods delivery.[2,3] Four types of mobile services are impacting consumer and traveler behavior. These include:

- **Mobility services** that assist users with routing, booking and payment of single and multimodal trips. This can include shared mobility, public transit apps, real-time information apps, taxi e-Hail and multimodal aggregators;
- **Courier network services** offering for-hire paid delivery for monetary compensation by employing an online application to connect couriers using private vehicles, bicycles, scooters or other equipment with light cargo;
- **Vehicle connectivity services** that provide vehicle diagnostic information and enable remote access and dispatch emergency services (e.g., accident and roadside assistance, unlocking a vehicle); and
- **Smart parking services** that deliver information on parking costs and availability. This includes "e-Parking" services to reserve and pay for parking and "e-Valet" services that connect vehicle owners to valet drivers to pick-up, park, charge or refuel, and return vehicles.

Additionally, real-time data analytics and algorithms are being used in a variety of ways to: (1) improve traveler experience, (2) enhance operations (such as managing crowdsourced and flexible routing), (3) provide predictive analytics to more accurately forecast and respond to demand, and (4) improve operational responses with natural or manmade hazards impacting usual transportation operations.[2,3,31]

Growing consumer use of digital services, coupled with real-time data analytics and algorithms, are creating vast amounts of data that will enable travelers, transportation providers and public agencies to make smarter, more intelligent and efficient transportation decisions.[2,3,31] While transportation service providers, such as public transit and shared mobility operators use a variety of data and data sources in their modeling and operations, big data and data sharing have the potential to enhance transportation planning and traveler services by empowering operators and policymakers to better understand the current state of the transportation network and more accurately identify services gaps and respond through immediate service adjustments and longer-term infrastructure improvements.[31]

Equally important and possibly more transformative, data collection and processing are key to deploying driverless vehicles in the MOD ecosystem.[3,4,31] Generally, data are used in three contexts to enable shared AVs (SAVs):

1. Machine learning algorithms predict outcomes based on large volumes of data;
2. Vehicle computers process detailed maps and other data for the entire street and adjacent environment; and
3. Vehicles communicate with other vehicles and infrastructure to share data.

As such, machine learning, big data management and other technologies are key enablers to connect travelers, goods, services and infrastructure as well as automating MOD vehicles.

8. MOD and Driverless Vehicles

In the future, automation could be one of the most transformative trends to impact MOD and the broader transportation ecosystem. Vehicle automation will likely result in fundamental changes to cities and the transportation ecosystem by altering the built environment, costs, commute patterns and modal choice. See Figure 4 for an overview of land use and built environments. In an AV future, transportation could change cities in four fundamental ways:

1. The density of urban centers could increase, as SAVs impact reliance on private vehicle ownership and use. Even privately owned AVs would no longer need to be parked in a city's highest valued real estate. Instead, these vehicles could self-drive and park away from residential, employment and other activity centers.[7] As such, auto-oriented land uses, such as parking and gas stations could be redeveloped into housing, offices and other land uses following the principles of the highest

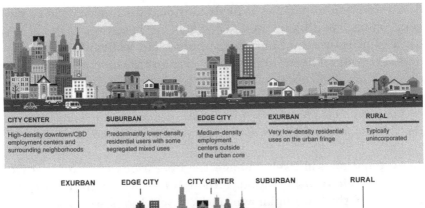

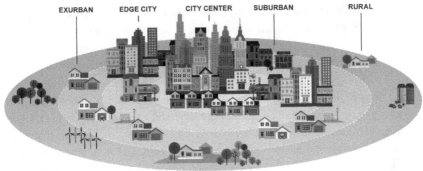

Fig. 4. Overview of land use and built environments.

Source: S. Shaheen, A. Cohen, B. Yelchuru and S. Sarkhili, *Mobility on Demand Operational Concept Report* (US Department of Transportation, Washington, 2017).

and best use. The four criteria guiding the highest and best use of real estate are: (1) legal permissibility, (2) physical possibility, (3) financial feasibility, and (4) maximum productivity.

2. Suburban and exurban areas are likely to expand, particularly in regions with high costs of living that lack affordable housing. With the growth of telework and AVs, fewer work days in the office longer commutes could become less of an impediment. Vehicle automation has the potential to transform commutes from lost driving time into productive hours that could be spent working, relaxing or sleeping. For these reasons, there could be a bifurcation in the built environment and modal choice, where urban centers become denser with greater adoption of shared modes, and suburban and exurban areas become more sprawling with continued reliance on privately owned (and now automated) vehicles.[3,7]

3. A reduction in parking is likely, although estimating the reduction amount is difficult and will likely be based on regional AV ownership rates, the built environment and walkability of a city, and the availability of high-quality public transportation and on-demand mobility options. Parking is a very expensive addition to most real estate projects, and the vast majority is unpaid with a limited return on the investment. A reduction in parking demand can free up land and capital to support other uses, such as increased density and public spaces.[32]

4. Automation will likely change the operations and competitiveness of public transportation, although the precise impacts (positive or negative) are unknown. Reduced vehicle ownership due to SAVs could result in changes in parking needs, particularly in urban centers. The repurposing of urban parking has the potential to create new opportunities for infill development and increased densities. While SAVs may compete with public transit ridership, infill development could also create higher densities to support additional public transit ridership and allow for the conversion of bus transit to rail transit in urban cores. Thus, concerns that the introduction of AVs could reduce demand for public transit and encourage increased vehicle use are real.[7] Just as AVs have the potential to reduce driving costs, automated transit vehicles have the opportunity to reduce operating costs and the potential to pass savings on to riders in the form of lower fares. Reduced operational costs and lower fares could make public transit more competitive with other modes and result in increased ridership.[7]

9. Conclusions

MOD is evolving and reshaping mobility. It is based on the principle that transportation is a commodity where modes have economic values that are distinguishable in terms of cost, journey time, wait time, number of connections, convenience and other attributes. MOD emphasizes the concept of transportation as a commodity, whereas MaaS focuses on integrating existing and innovative mobility services into one single digital platform where customers purchase mobility service packages tailored to their individual needs. The MOD ecosystem is comprised of numerous public and private stakeholders, such as mobility and delivery service providers that influence the supply and demand of the transportation system. Information technology, such as real-time data and smartphone apps, coupled with innovative business models are key MOD enablers.

While the impacts of vehicle automation on MOD are uncertain, the convergence of mobility services, shared modes, electrification and automation could transform how people travel, how cities are designed and built, and the ways in which consumers access goods and services. As emerging technologies become more mainstream policymakers will need to rethink traditional notions of access, mobility and auto mobility. While increased use of multi-passenger vehicles and optimized routing could greatly increase efficiency, lower costs and improved accessibility could offset these gains due to induced demand, which could result in a net increase in energy use and emissions. Moving forward, it is critical to balance public goals, commercial interests and technological innovation to harness and maximize the social and environmental benefits of next generation mobility systems.

References

1. S. Shaheen, A. Cohen and I. Zohdy, *Shared Mobility: Current Practices and Guiding Principles* (US Department of Transportation, Washington, 2016).
2. S. Shaheen, A. Cohen, I. Zohdy and B. Kock, *Smartphone Applications to Influence Travel Choices: Practice* (US Department of Transportation, Washington, 2016).
3. S. Shaheen, A. Cohen, B. Yelchuru and S. Sarkhili, *Mobility on Demand Operational Concept Report* (US Department of Transportation, Washington, 2017).
4. S. Shaheen, A. Cohen, M. Dowd and R. Davis, *A Framework for Integrating Transportation into Smart Cities: State of the Practice in 20 US Cities* (Mineta Transportation Institute, San Jose, 2019).
5. S. Shaheen, A. Cohen and E. Martin, *The US Department of Transportation's and the Federal Transit Administration's Mobility on Demand Sandbox* (US Department of Transportation, Washington, 2017).
6. S. Shaheen and A. Cohen, *Mobility on Demand: Three Key Components*, 27 February 2018. https://www.move-forward.com/mobility-on-demand-three-key-components/.
7. S. Shaheen and A. Cohen, "Is it time for a public transit renaissance? Navigating travel behavior, technology, and business model shifts in a brave new world", *Journal of Public Transportation*, 2018, **21** pp. 67–81. See: https://scholarcommons.usf.edu/jpt/vol21/iss1/8/.
8. A. Durand, L. Harms, S. Hoogendoorn-Lanser and T. Zijlstra, *Mobility-as-a-Service and Changes in Travel Preferences and Travel Behaviour: A Literature Review* (KiM Netherlands Institute for Transport Policy Analysis, Amsterdam, 2018).

9. S. Hietanen, "Mobility as a service: The New Transport Paradigm", *ITS & Transport Management Supplement. Eurotransport*, 2014, pp. 2–4. http://fsr.eui.eu/wp-content/uploads/150309-1-Hietanen-1.pdf.

10. R. Utriainen and M. Pöllänen, (2017). "Review on mobility as a service in scientific literature". *Conference Proceedings, 1st International Conference on Mobility as a Service (IcoMaaS)*, Tampere University of Technology, Tampere, 28 to 29 November 2017, pp. 141–155.

11. A. Melis, S. Mirri, C. Prandi, M. Prandini, P. Salomoni and F. Callegati, "Integrating personalized and accessible itineraries in MaaS ecosystems through microservices", *Mobile Networks and Applications*, 2018, pp. 167–176.

12. A. Melis, M. Prandini, L. Sartori and F. Callegati, "Public transportation, IoT, trust and urban habits", in *Public Transportation, IoT, Trust and Urban Habits* (Springer International Publishing AG, Cham, 2016).

13. K. Pangbourne, M. Mladenovic, D. Stead and D. Milakis, "The case of Mobility as a Service: a critical reflection on challenges for urban transport and mobility governance", in *Governance of the Smart Mobility Transition* (Bingley: Emerald Group Publishing, Bingley, 2017), pp. 33–48.

14. G. Ambrosino, J. Nelson, M. Boero and I. Pettinelli, "Enabling intermodal urban transport through complementary services: from flexible mobility services to the shared use mobility agency: workshop 4. Developing intermodal transport systems", *Research In Transportation Economics*, 2016, pp. 179–184.

15. D. Hensher, "Future bus transport contracts under a mobility as a service (MaaS) regime in the digital age: are they likely to change?" *Transportation Research Part A: Policy and Practice*, 2017, pp. 86–96.

16. M. Kamargianni, W. Li, M. Matyas and A. Schäfer, "A critical review of new mobility services for urban transport", *Transportation Research Procedia 1*, 2016, pp. 3294–3303.

17. M. Kamargianni and M. Matyas, "The Business Ecosystem of Mobility-as-a-Service", *Transportation Research Record: Journal of the Transportation Research Board*, 2017.

18. H. Ozaki, "Technical standardization of ITS and Asian initiatives for intelligent mobility", *IATSS Research*, 2018, pp. 72–75.

19. A. Melis, "Responding to climate change: lessons from an Australian hotspot", *Urban Policy and Research*, 2017, pp. 114–115.

20. T. Hilgert, M. Kagerbauer, T. Schuster and C. Becker (2016). "Optimization of individual travel behavior through customized mobility services and their effects on travel demand and transportation systems", *Transportation Research Procedia*, 2016, pp. 58–69.

21. P. Jittrapirom, V. Caiati, A.-M. Feneri, S. Ebrahimigharehbaghi, M. Alonso González and J. Narayan, "Mobility as a service: a critical review of definitions, assessments of schemes, and key challenges", *Urban Planning*, 2017.

22. J. Sochor, H. Arby, M. Karlsson and S. Sarasini, "A topological approach to Mobility as a Service: A proposed tool for understanding requirements and effects, and for aiding the integration of societal goals", *Research in Transportation Business & Management*, 2018, pp. 3–14.
23. S. Shaheen and A. Cohen, "Mobility on Demand (MOD) and Mobility as a Service (MaaS). How are they similar and different?" *Move Forward*, 7 March 2019. https://www.move-forward.com/mobility-on-demand-mod-and-mobility-as-a-service-maas-how-are-they-similar-and-different/.
24. S. Shaheen and A. Cohen, *Planning for Shared Mobility* (The American Planning Association, Chicago, 2016).
25. S. Shaheen, E. Martin, N. Chan, A. Cohen and M. Pogodzinski, *Public Bikesharing in North America During A Period of Rapid Expansion: Understanding Business Models, Industry Trends, and User Impacts* (Mineta Transportation Institute, San Jose, 2014).
26. S. Shaheen and A. Cohen, "Shared micromobility policy toolkit", March 2019. http://doi:10.7922/G2TH8JW7.
27. S. Shaheen, A. Cohen and D. Roberts, "Carsharing in North America: market growth, current developments, and future potential", *Transportation Research Record*, 2005.
28. S. Shaheen, A. Cohen and E. Farrar, *The Potential Societal Barriers of Urban Air Mobility (UAM)* (National Aeronautics and Space Administration, Washington, 2018).
29. National Aeronautics and Space Administration, "NASA embraces urban air mobility, calls for market study", National Aeronautics and Space Administration. https://www.nasa.gov/aero/nasa-embraces-urban-air-mobility.
30. S. Shaheen and A. Cohen, "Seven goods delivery innovations: transforming transportation & consumer behavior", *Move Forward*, 12 December 2018. https://www.move-forward.com/seven-goods-delivery-innovations-transforming-transportation-consumer-behavior/.
31. S. Shaheen and A. Cohen, "Big data, automation, and the future of transportation", *Meeting of the Minds*, 25 July 2017. https://meetingoftheminds.org/big-data-automation-future-transportation-22106.
32. S. Shaheen and A. Cohen, "Managing the transition to shared automated vehicles: building today while designing for tomorrow", *Meeting of the Minds*, 21 May 2018. https://meetingoftheminds.org/managing-the-transition-to-shared-automated-vehicles-building-today-while-designing-for-tomorrow-2-27094.

Chapter 5

Urban Energy Systems Design: A Research Outline

Perry Pei-Ju Yang

Eco Urban Laboratory, School of City and Regional Planning and School of Architecture, Georgia Institute of Technology, Atlanta, USA
perry.yang@design.gatech.edu

1. Introduction

The energy efficiency of a society depends on the performance of a city operating as an integrated system. We are only beginning to understand how high performance buildings, transportation systems, renewable energy systems, distributed generation and other components of a modern city can work together as complex urban systems. Unfortunately, we are only beginning to understand how such systems can operate and much research remains too immature to provide guidance essential to policymakers. The research that is being done is often difficult to locate since it is published in a very diverse set of journals. Research on how urban energy systems can be designed to be more efficient and resilient is urgently needed. This chapter points to one of the most promising research issues of our time.

There is growing recognition of the key role that cities play in the world's energy future. Cities contribute to 67% of the global primary energy demand, and 71% of energy-related carbon dioxide emissions.[1] Nearly half of the world's population (6.5 billion people) live in cities. By 2040, an extra 1.7 billion people will be added mostly to urban areas in developing economies, increasing global energy demand by more than a quarter.[1,2] About 3.2 billion people will be added to cities by 2050.[3] Assuming that there will be 9 billion people living on the planet then, and that the entire global population will adopt the current living standard of a US resident, which takes 1.1361×10^8

tw of energy to sustain, the world will need about 102 tw of energy in 2050. Given that the main population growth will occur in Asia and Africa,[3] in which relatively low per capita energy use will be needed, conservative estimates of global energy needs by considering country differences will be around 28–35 TW in 2050, more than double the global energy demand today.[4]

Models of urban systems design need to consider both technical issues of optimization and many social and behavioral issues that affect urban performance. Energy system designs must be able to meet the complex needs of urban communities as well as the many new risks amplified by climate change such as heat waves, flooding and the urban heat island. Modern cities, however, were not necessarily designed with energy efficiency and resiliency in mind. Urban forms were often produced according to principles such as land development and zoning regulations, rather than resilience to climate change. Research on urban energy systems is emerging that sees cities as complex systems, in which energy, material, water flows and human movement are central. This perspective gives rise to a critical question: what are the transformative strategies of urban energy systems to make cities more efficient in performance, more renewable in resource management, and more resilient in their systems behavior?

The chapter suggests five issues required to outline a research agenda of urban energy that are keys to developing a resilient and sustainable urban future:

1. Energy and urban form: How does urban form as a physical spatial structure perform with regard to energy use?
2. A system of systems: What is the cross-sectorial interdependency between energy, water and food systems?
3. Context matters: What are the multi-scale relations in the urban environment concerning energy performance?
4. Defining system boundaries: How are system boundaries defined for urban energy accounting?
5. Design as a transformative approach: What is a transformative approach to promoting system efficiency and resiliency in urban energy systems?

2. Energy and Urban Form

The relationship between energy and cities was explored in the early 1980s by Newman and Kenworthy. They found that the average gasoline consumption in US cities was nearly twice as high as in Australian cities, four times higher than in European cities and ten times higher than in Asian cities.[5] Low-density urban environments that exist in many American cities produce long distance commuting that consumes more gasoline than the high-density urban settings in many Asian cities that rely on public transportation. Figure 1 shows the strong relationship between urban density and per capita gasoline use.

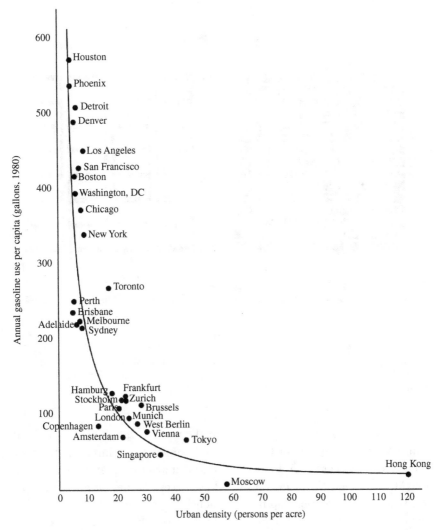

Fig. 1. Gasoline use per capita vs. urban density.[5]

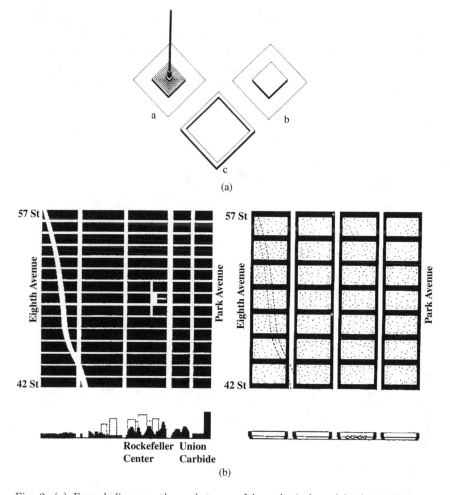

Fig. 2. (a) Fresnel diagram, the archetypes of hypothetical models that include: a. tower, b. pavilion, and c. courtyard; and (b) Manhattan proposal: the tower and podium or pavilion typologies (left) and courtyard typology (right).[6]

The above research only considered the transportation sector, without counting the energy use of the built environment. The internal structure of cities or physical urban form has a major impact on energy use. Martin and March raised the question of "what makes the best use of land" by constructing the relationship between density, spatial typology and performance.[6] The principle of the Fresnel Diagram (see Fig. 2(a)) addresses the effectiveness of distributing built volume on the perimeter of a site. That

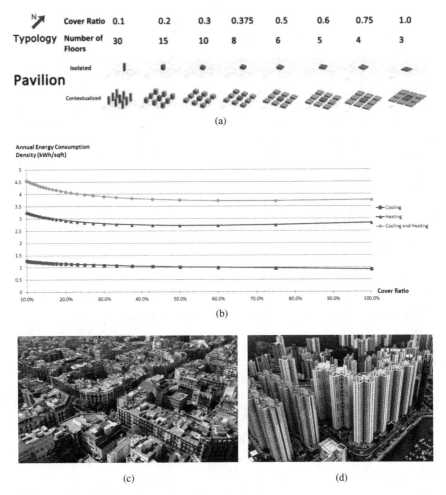

Fig. 3. The relationship between urban form and energy: given the same building density, cover ratio increases from 10% to 100% (a); energy consumption per square feet decreases according to cover ratio change (b); and two same urban forms: Hong Kong for low cover ratio and Barcelona for high cover ratio (c). (Quan *et.al.*, 2014).[10]

means that given the same building density, a variety of urban forms and building typologies can be distributed on an urban site, and would potentially generate different patterns of energy flow as a consequence of the environmental effects of urban form, such as daylighting, ventilation, solar availability and building energy. The relationship between density, urban form typology and energy is to be explored for optimizing system performance. Martin and March developed a provocative proposition for Manhattan as a test bed to explore alternative urban forms and building archetypes as hypothetical models (see Fig. 2(b)). The current Manhattan urban form that is mainly composed of the tower and podium typology is to be replaced by the courtyard typology that reduces the average building height of 21 stories to seven stories while maintaining the same density.

The alternative pattern of built form and open space would produce different energy and ecological effects. Urban form matters in energy performance. Given the same building density, different urban form and building typology may generate different energy performance due to the variety of building shapes, building design and HVAC systems that would perform with different building energy uses. The spacing in-between buildings and urban canyon effects would affect the energy performance too. The following test bed shows that when the cover ratio increases from 10% to 100%, the energy consumption per square feet decreases accordingly. Urban form in different cultural and climate contexts matters. Urban designers manipulate the deployment of urban density and decisions of building typologies, and this impacts the urban energy agenda.

3. A System of Systems

Urban energy deals with a complex system that is composed of the subsystems of cities such as energy, materials, water, transportation and information. What is the cross-sectorial interdependency among energy, water and food systems? The problems of urban energy rely on integration of systems through understanding the interdependencies across sectorial boundaries in cities.

A near zero energy district project was proposed using Disney Shanghai's surrounding site as a test bed. It is an attempt to integrate infrastructural systems, driven by resources renewability, system resiliency and a place-based decentralization to meet the objective of designing a near zero energy urban system. The near zero system is composed of infrastructures including power grids, material life cycle loops, water supply and treatment systems and building systems, in which interdependent relations among subsystems are to be constructed (see Fig. 4). The use of decentralized renewable energy systems aims to minimize the input and output of energy, water and resources from the environment. Decentralization also leads to decentralized decision making, weakening the control exercised by institutions that traditionally managed centralized energy, water

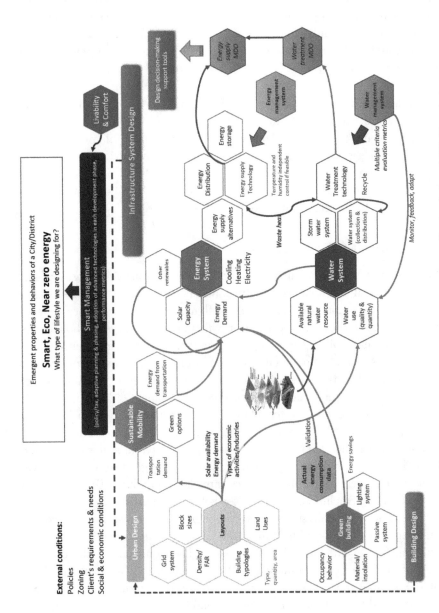

Fig. 4. The System of Systems for the near zero energy district design: 10KM2 project.

and waste management systems. The system depends on a network of local regulatory organizations responsible for individual systems and integration. Smart local management can make these systems smarter, more sustainable, and more resilient.

For architects and urban designers who are shaping the physical built environment, the following questions are the most crucial:

1. What are the impacts of renewable energy to the place-based infrastructure and urban form?
2. How would the decentralization strategies for renewable infrastructure derive a set of new organizational principles for designing cities and urban spaces?
3. How would cities, neighborhoods and building block systems be reorganized to accommodate the new infrastructure?

4. Context Matters

What are the cross-scale interactions and dynamics of urban energy systems from individual components, buildings and neighborhoods to cities? What property would emerge when energy systems scale up from finer-scale to broader-scale territories, in which energy flows across the system boundaries from a singular building system to a complex urban environment? There exists extensive literature on energy performance and simulation of building systems, in which the focus of urban and community-scale building energy modeling is emerging that needs further development.[7,8] The multi-scale model for urban energy system design needs greater effort both in theory and experiment. Urban systems contain processes and flows of energy, material and information that operate across different spatial scales and system boundaries in the built environment, in which their processes co-evolve over time. Urban energy modeling involves three key factors: (1) Urban physical context, (2) Microclimate and (3) User behavioral patterns[7]:

(1) The urban physical context is defined by interactions between a building and other buildings and obstructions. As solar radiation on the building facade influences building energy use significantly, the obstruction of sunlight by surrounding buildings, trees and other obstructions plays an important role in building energy use. (2) Urban energy modeling should consider interactions between buildings and the microclimates around them. Urban heat island (UHI) effect, the phenomenon of higher air temperatures in built areas as compared to the air temperatures in the

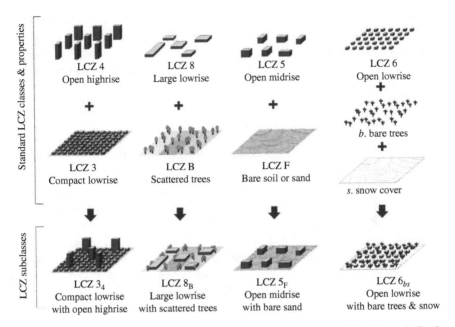

Fig. 5. Definition of local climate zones that are combinations of "built" and "land cover" types.[9]

surrounding rural countryside, is caused by urban form and landscape related factors. The effects of greater absorption of solar radiation and retention of infrared radiation in urban street canyons modify urban microclimate and then affect urban building energy. Steward and Oke proposed a detailed classification system for urban form to improve the UHI studies. They categorized urban areas into "local climate zones", i.e., regions of homogenous surface cover, material, structure and human activity, with the area radius of hundreds of meters to a few kilometers (see Fig. 5).[9] (3) Interactions between buildings and occupants are significant. The occupancy pattern, including the density and behavior, could lead to variations of building energy use in the same building.

These three factors induce complexity and uncertainty in urban energy problems. Properties of complexity emerge when the system boundary goes beyond a building and engages other buildings, infrastructures and the surrounding environment. Due to uncertainties of infiltration rates across building envelops and occupant behavioral patterns, the impact of urban microclimate on building energy use is hard to predict.[7,8] To conduct urban energy system research, we need to develop a multi-scale model of energy flow and patterns that can be observed in the built environment at different spatial scales and at hierarchical levels, and to articulate them systematically so as "to optimize model complexity and to reduce uncertainty".[11]

5. Defining System Boundaries

As it is possible "to fail to see the forest for the trees, it is possible to fail to see the city for the buildings".[12] A comprehensive assessment method of urban energy is needed. How much is the energy use of an urban district of a city and what are the reduction strategies? The answer surprisingly varies according to the system boundary, in other words, a conceptual boundary defined in a physical or functional space, which encloses all components that form the urban system.[13] How system boundaries are defined for urban energy accounting is therefore very important since the performance results may change enormously as a function of the alternative spatial, temporal, sectorial and functional system boundaries chosen.

Steinberger and Weisz summarized four energy accounting methods: the direct final energy account, regional energy metabolism, regional economic activity and energy input-output model.[13] Four types of emissions inventories were proposed: direct, responsible, deemed and logistic (see Fig. 6[14]). Through these methods, energy use and carbon emissions can be estimated by measuring the final energy used or consumed within the city boundary, and the energy used for producing goods and services that may be consumed in the city or be exported from the city and consumed in other cities or hinterlands, etc.

However, they are mainly conducted based on the top-down approach. For the urban, community and neighborhood scale territories, how do we define system boundaries of urban energy systems, in which energy flows transcend traditional boundaries such as geographical–territorial, social-demographic and political-institutional borders? How does energy function and flow in cities across different levels of urban form, from buildings, street blocks, urban districts and communities to cities, and how do we measure it? Traditional administrative boundaries and political territories that define "cities" have to be complemented by an "urban system" perspective, in which urban problems are to be understood through "functional dimensions" that transcend territorial and administrative boundaries.[12,13]

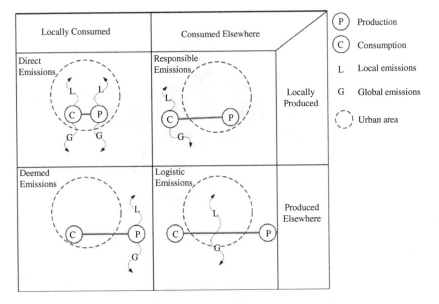

Fig. 6.   Four typologies of carbon emissions.[14]

6.   Design as a Transformative Approach

What is the transformative approach to promoting the system efficiency and resiliency of urban energy systems? How does urban design as an intervention transform and change urban energy systems by promoting the system efficiency and resiliency?

Sustainable urban design and planning are discussed many times in recent literature. However, these discussions focus mostly on normative criteria or policy-oriented tools. System approaches to connecting data analytics and urban design are largely unexplored in urban design practice. The distinction between traditional urban design and urban systems design has to be made. Traditional urban design tends to be deterministic, while a systems approach to urban design deals with energy and ecological processes that are sometimes stochastic. The design approach to an urban energy system is based on the measures of system performances of urban form, energy and material life cycle. They are situated in cross-scale contexts of hierarchical complex systems of cities, and should be considered in a temporal dimension, system threshold and stochastic process to manage urban change and make sustainable progress. Individual decisions of

Energy Efficiency

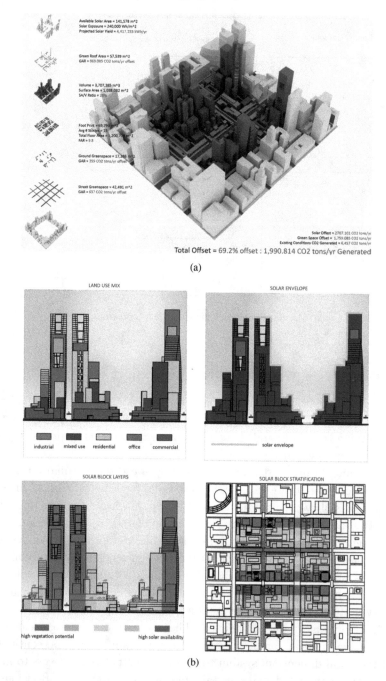

(a)

(b)

Fig. 7. Example of how an urban design scenario promotes energy performance of an urban district.[15]

urban design are also made in the context of a regulatory and policy structure. A good urban energy policy would encourage sustainable decisions that maximize social as well as individual benefits.

Urban energy research requires an understanding of how urban systems function and perform as regard to energy, a positive approach towards "science for design" that is based on systems science. The question of how urban energy systems can be enhanced would need a transformative approach. Urban systems design as an interventional approach brings in the question of "design in science", i.e., how urban energy systems can be changed through design. It is a normative question about how good city forms are designed, and a forward-looking perspective to deal with how urban systems can be designed to promote a city's efficiency and resiliency.

The following test bed shows urban design as an instrument to synthesize complex and dynamic factors for deriving sustainable urban systems scenarios. A design model is projected based on an existing urban district at the Chicago Loop. There is a long-range plan for transforming the current urban block to be more energy efficient and resilient. Some guiding principles of urban design are introduced in a process of reconfiguring urban form, building typologies and their relationship. For example, better spacing between buildings was tested for maximizing solar capture and enhancing air ventilation. Multiple levels of vegetation were introduced to roof tops, void decks and grounds for cooling down the temperature. Given the same building density by redesigning a current urban form structure, the urban energy performance can be enhanced by 63.7% through adding the reduction of energy consumption and the increase of renewable energy gain (see Fig. 7).

7. Conclusions

Sustainable approaches to urban systems research are emerging in the recent literature concerning industrial ecology, environmental system engineering and city planning. However, these approaches focus mostly on non-spatial aspects. Cities at the systems level are not well understood,[16] in which operational approaches to connecting urban design and energy performance and other ecological analyses are largely unexplored.

Sustainable urbanism is about infrastructural network and flows, in which both spatial and temporal dimensions mediate flows and forms of the environment. Urban designers and systems scientists offer perspectives that are to be coordinated for urban energy problems that are complex and full of uncertainty. Urban energy problems require a systems perspective to comprehend how cities function, as well as design as a transformative approach to promote urban energy efficiency and system resiliency. Urban systems design is a tool for synthesizing complex factors and for speculating about the future city model.

This chapter outlined promising research issues for the growing urban world and proposed possible approaches that are critical for meeting the challenges of an urban energy future. One critical dimension missing in this chapter is how social behavioral principles enhance urban energy performance. Beyond the human behavioral patterns, how does urban energy systems research incorporate human interactions? To what extent do the social and behavioral changes promote an energy agenda, particularly in the context of emerging technologies, the internet of things and the smart city movement? The social impact of an energy efficient city and its cross-scale effects from a building to the levels of campus, urban neighborhood and landscape, to the city itself are still to be articulated.

References

1. IEA, "World energy outlook", International Energy Agency, 2008. https://www.iea.org/weo/.
2. The International Energy Agency, World Energy Outlook 2018; https://www.iea.org/weo/Ibid.
3. UN, "World population prospect 2019: highlights", United Nations Department of Economic and Social Affairs, 2019.
4. D.G. Nocera, "On the future of global energy", American Academy of Arts and Sciences, 2006.
5. P.W.G Newman and J.R. Kenworthy, "Gasoline consumption and cities", *Journal of the American Planning Association* **55** (1989) 24–37. https://doi.org/10.1080/01944368908975398.
6. L. Martin and L. March, *Urban Space and Structures* (Cambridge University Press, London, 1972).
7. S.J. Quan, Q. Li, G. Augenbroe, J. Brown and P.P.J. Yang "Urban data and building energy modeling: a GIS-based urban building energy modeling system using the urban-EPC engine", in *Planning Support Systems and Smart Cities* (Springer, 2015), pp. 447–469.
8. C.F. Reinhart and C.C. Davila, "Urban building energy modeling: a review of a nascent field", *Building and Environment* **97** (2016) 196–202.
9. D.I. Stewart and T.R. Oke "Local climate zones for urban temperature studies", *Bulletin of the American Meteorological Society* **93** (2012) 1879–1900.

10. S. J. Quan, A. Economou, T. Grasl and P. P-J. Yang "Computing energy performance of building density, shape and typology in urban context", *Energy Procedia*, **61** (2014) 1602–1605.
11. V. Grimm, E. Revilla, U. Berger, F. Jeltsch, W.M. Mooij, S.F. Railsback, H.H. Thulke, J. Weiner, T. Wiegand and D.L. DeAngelis, "Pattern-oriented modeling of agent-based complex systems: lessons from ecology", *Science* **310** (2005) 987.
12. A. Grubler and D. Fisk, "Introduction and Overview", in *Energizing Sustainable Cities: Assessing Urban Energy* (EarthScan, 2013).
13. J. Steinberger and H. Weisz, "City walls and urban hinterlands: the importance of system boundaries", in *Energizing Sustainable Cities: Assessing Urban Energy* (EarthScan, 2013).
14. B.K. Sovacool and M.A. Brown, "Twelve metropolitan carbon footprints: A preliminary comparative global assessment", *Energy Policy* **38** (2010) 4856–4869.
15. P.P.J. Yang (ed.), "Global benchmarking for low carbon urban design", Ecological Urbanism Studio Report, Georgia Institute of Technology, 2011.
16. T.E. Graedel and B.R. Allenby *Industrial Ecology and Sustainable Engineering* (Prentice Hall, 2010).

Chapter 6

Energy and Aviation

Gaudy M. Bezos-O'Connor[*,‡], Jeann C. Yu[†,§], Doug Christensen[†,¶],
Al Sipe[†,‖], Tia Benson Tolle[†,**], Tim Rahmes[†,††] and David Paisley[†,‡‡]

NASA Langley Research Center Hampton, VA 23681, USA
†*The Boeing Company, Seattle, WA 98124, USA*
‡*gaudy.m.bezos-oconnor@nasa.gov*
§*jeannecyu@hotmail.com*
¶*douglas.p.christensen@boeing.com*
‖*alvin.l.sipe@boeing.com*
****tia.h.bensontolle@boeing.com*
††*Timothy.F.Rahmes@boeing.com*
‡‡*david.j.paisley@boeing.com*

1. Introduction

This chapter explores the regulatory and technical drivers that can shape the future of energy efficient air travel.[a] Aviation, which is growing rapidly, contributed 2.4% of global energy-related carbon dioxide emissions in 2018.[1] Aviation emission have increased by 26% over the past five years; 4/5 of these emissions have come from passenger transport. Driven by the increasingly global nature of business and leisure travel, demand is projected to increase by 3.8% annually and reach 17 trillion passenger kilometers by 2040, more than triple the distance traveled in 2010.[b] The increase

---

[a]Though aircraft-induced clouds or contrails are responsible for about half of the greenhouse effect (radiative forcing) created by aircrafts, this chapter will only address direct energy use by aircrafts and their associated carbon dioxide emissions. See https://www.nature.com/articles/s41467-018-04068-0#Tab1.

[b]International Energy Agency, "World energy outlook 2017", Organisation for Economic Co-operation and Development/International Energy Agency, 2017. https://www.iea.org/weo2017/.

in demand can be partially offset by gains in efficiency. The aircraft fleet in service today is over 80% more efficient per passenger kilometer than the aircrafts of the 1960s.[c]

The International Civil Aviation Organization (ICAO) monitors technological advances and conducts fleet-level aviation efficiency analyses to update forecasts with regard to the achievement of fuel burn goals, emissions and noise reduction goals, and adherence to established regulatory standards. These are updated every three years. The organization has established a target goal of 2% annual improvements in fuel efficiency and Carbon Neutral Growth from 2020 onwards.[d]

Figure 1 shows the fuel burn trend for international aviation from 2005 to 2045, which has also been extrapolated to 2050. This analysis takes into account improvements in operational and air traffic management as well as the contribution of available aircraft technology. The individual contributions were calculated to be .0.98% for technology improvements and 0.39% for operational improvements. Uncertainties in these forecasts are large, larger than the range of potential contributions from technological and operational improvements.[e] The projected long-term fuel efficiency gains shown in this analysis, however, are 1.37% per year in the best-case scenario — short of the ICAO's goal.

But much greater gains in efficiency are possible. The National Aeronautics and Space Administration (NASA) has set a goal of increasing aircraft fuel efficiency to 60% above its 2005 levels by 2035. These gains can be achieved through technological advances in engine designs, aerodynamic performance improvements as well as operational optimizations, including shortening flight distances through better routing, terminal area (taxi-out, taxi-in) and flight path (climb-out, cruise, descent) optimization[f] (see Fig. 2). The technologies needed to meet these ambitious targets will be discussed in detail in later sections of this chapter.

---

[c]Air Transport Action Group, "Facts and figures", Air Transport Action Group. https://www.atag.org.

[d]International Civil Aviation Organization, "Technology and standards", International Civil Aviation Organization. https://www.icao.int/environmental-protection/ Documents/ICAOEnvironmental_Brochure-1UP_Final.pdf.

[e]International Civil Aviation Organization, "2019 environmental report", International Civil Aviation Organization, 2019. https://www.icao.int/environmental-protection/ Documents/ICAO-ENV-Report2019-F1-WEB%20%281%29.pdf.

[f]"United States aviation greenhouse emissions reduction plan", submitted to International Civil Aviation Organization, June 2015. https://crp.trb.org/acrp0267/ united-states-aviation-greenhouse-gas-emissions-reduction-plan/.

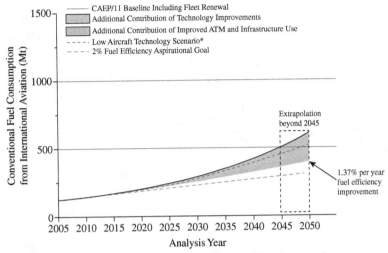

Fig. 1.   International aviation fuel burn from 2005 to 2050.

Note: Results were modelled for 2005, 2006, 2015, 2025, 2035, and 2045 then extrapolated to 2050.

BJ: Business Jet; RJ: Regional Jet; SA: Single Aisle; TA: Twin Aisle.

Source: ICAO Environmental Report, 2019.

*Dashed line in technology contribution silver represents the "Low Aircraft Technology Scenario".

The projected lifecycle carbon dioxide emission impacts shown in Fig. 2 for the aggressive system improvement scenario use the Federal Aviation Administration's (FAA) modeling and analysis tools.[2] Each "wedge" represents the amount of carbon dioxide reduced from a baseline scenario due to an improvement category. The models do not attempt to reflect long-term structural changes in the global transportation system that may be driven by such factors like the growth of megacities and changes in demand for moving/transporting people and freight.

1.1.   *Industry Strategy*

The forecast above confirms that technology alone will not meet the ICAO's fuel burn goal and thus a substantial increase in investment in aircraft technology is urgently required. As some countries were contemplating carbon dioxide standards for their aviation fleets, the industry

Energy Efficiency

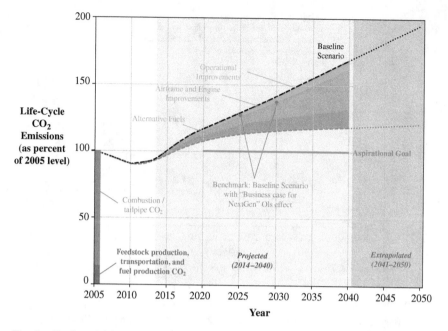

Fig. 2. Projected lifecycle carbon dioxide emissions impacts for the aggressive system improvement scenario.

Source: https://www.icao.int/environmental-protection/Lists/ActionPlan/Attachments/30/UnitedStates_Action_Plan-2015.pdf.

sought support for global standards due to the international nature of air transport, where a myriad of regional standards would cause undue complexity. The global industry came together with the ICAO to set aggressive carbon dioxide reduction goals, and develop industry fuel burn and carbon dioxide certification standards for new aircrafts.

In October 2013 the ICAO adopted a Carbon Offsetting and Reduction Scheme for International Aviation (CORSIA) designed to:

- "Stabilize net aviation carbon dioxide emissions at 2020 levels through Carbon Neutral Growth", and
- "Reduce the aviation industry's net carbon dioxide emissions to 50% of what they were in 2005 by 2050".

From 2020 onwards, emissions would be reduced through improved aircraft technology, operational improvements, sustainable aviation fuels, and global market-based measure as shown in Fig. 3.

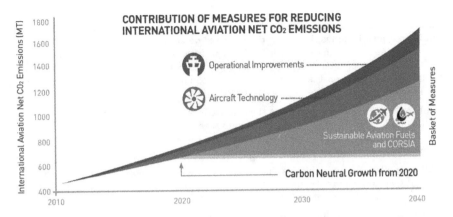

Fig. 3. Aviation industry global transport climate action framework goals.

Carbon offsetting or the reduction of carbon dioxide emissions can be achieved through the acquisition and retirement of emissions units from the global carbon market by aircraft operators.[g]

The approach was designed to be flexible. As of January 2018, there are 73 countries participating in the scheme, which accounts for 76% of international aviation passenger kilometers.[h] CORSIA covers only international travel, thus domestic flights that contribute over 40% of aviation fuel consumption are exempt.[3] It also exempts small national carriers with less than 0.5% of the 2018 passenger kilometer market share.

In March 2017, the Committee on Aviation Environmental Protection, at its 10th meeting (CAEP/10), formally adopted the first-ever international standards to regulate aircraft carbon dioxide emissions for all new commercial and business aircrafts starting in 1 January 2028, as shown in Table 1. There will be a transition period for modified aircrafts starting in 2023.

The ICAO's certification standard will apply to new aircraft type designs from 2020 and to aircraft type designs already in-production as of 2023. However, aircrafts that are still in-production by 2028 and do not

[g]International Civil Aviation Organization, "Carbon offsetting and reduction scheme for international aviation", International Civil Aviation Organization. https://www.icao.int/environmental-protection/Documents/ICAOEnvironmental_Brochure-1UP_Final.pdf.

[h]International Civil Aviation Organization , "CORSIA states for chapter 3 state pairs", International Civil Aviation Organization, 2020. https://www.icao.int/environmental-protection/CORSIA/Documents/CORSIA_States_for_Chapter3_State_Pairs_Jul2020.pdf.

Table 1. The ICAO's carbon dioxide standard for aircrafts.

Subsonic jet airplanes, including their derived versions, of greater than 5,700 kg maximum take-off mass, for which the application for a type certificate was submitted on or after 1 January 2020, except for airplanes of less than or equal to 60,000 kg maximum take-off mass with a maximum passenger seating capacity of 19 seats or less;

Subsonic jet airplanes, including their derived versions, of greater than 5,700 kg and less than or equal to 60,000 kg maximum take-off mass with a maximum passenger seating capacity of 19 seats or less, for which the application for a type certificate is submitted on or after 1 January 2023;

All propeller-driven airplanes, including their derived versions, of greater than 8,618 kg maximum take-off mass, for which the application for a type certificate was submitted on or after 1 January 2020;

Derived versions of non-carbon-dioxide-certified subsonic jet airplanes of greater than 5,700 kg maximum take-off mass, for which the application for certification of the change in type design was submitted on or after 1 January 2023;

Derived versions of non-carbon-dioxide-certified propeller-driven airplanes of greater than 8,618 kg maximum certificated take-off mass, for which the application for certification of the change in type design was submitted on or after 1 January 2023;

Individual non-carbon-dioxide-certified subsonic jet airplanes of greater than 5,700 kg maximum certificated take-off mass, for which a certificate of airworthiness was first issued on or after 1 January 2028;

and

Individual non-carbon-dioxide-certified propeller-driven airplanes of greater than 8,618 kg maximum certificated take-off mass, for which a certificate of airworthiness was first issued on or after 1 January 2028.

Source: ICAO's Carbon Offsetting and Reduction Scheme for International Aviation. https://www.icao.int/environmental-protection/Documents/ICAOEnvironmental_Brochure-1UP_Final.pdf.

meet the standard will no longer be able to be produced unless their designs are sufficiently modified.

1.2. *Federal agency goals*

The United States' National Aeronautics and Space Administration (NASA) and the Federal Aviation Administration (FAA) lead efforts to transform the aviation industry by dramatically reducing its environmental impact and improving its efficiency. In 2009, NASA's Aeronautics

TECHNOLOGY BENEFITS	TECHNOLOGY GENERATIONS (Technology Readiness Level = 5-6)		
	Near Term 2015–2025	Mid Term 2025–2035	Far Term beyond 2035
Noise (cum below Stage 4)	22–32 dB	32–42 dB	42–52 dB
LTO No$_x$ Emissions (below CAEP 6)	70–75%	80%	>80%
Cruise No$_x$ Emissions (rel. to 2005 best in class)	65–70%	80%	>80%
Aircraft Fuel/Energy Consumption (rel. to 2005 best in class)	40–50%	50–60%	60–80%

Fig. 4.   NASA targets for subsonic transport systems.

Source: National Aeronautics and Space Administration, "NASA aeronautics: strategic implementation plan–2017", update. NP-2017-01-2352-HQ, National Aeronautics and Space Administration. https://www.nasa.gov/sites/default/files/atoms/files/sip-2017-03-23-17-high.pdf.

Research Mission Directorate established the Subsonic Transport Metrics, with aggressive targets for aviation technology development as shown in Fig. 4. These goals challenged industry norms and its sister federal laboratories to develop or mature technologies to reduce aircraft energy consumption. The goals (including other environmental goals) aim to reduce fuel burn, Cruise, Landing and Take-off Nitrous Oxides emissions relative to the CAEP 6[i] guidelines, and cumulative noise relative to the Stage 4 guidelines within the Entry into Service (EIS) timeframes.[j] The FAA's Continuous Lower Energy, Emissions and Noise Program, which was created in 2010, adopted the NASA metrics defined for the near term timeframe (2015–2025) and has successfully implemented two 5-year Public-Private partnership cycles with the aviation industry and NASA, to accelerate and expedite integration of advance noise, emissions, fuel burn and advance sustainable alternative jet fuels technology into the fleet within a 5-year development window.

---

[i]International Civil Aviation Organization (ICAO) Committee on Aviation and Environmental Protection (CAEP) 6/NOx Standards. https://www.icao.int/Meetings/ENVSymposium/Presentations/Neil Dickson Session 5v2.pdf.
[j]National Aeronautics and Space Administration (NASA), "NASA targeted improvements in subsonic transport system-level metrics", National Aeronautics and Space Administration. https://www.nasa.gov/sites/default/files/atoms/files/sip-2019-v7-web.pdf.

2. Today's Aviation Fleet Composition

The first century of aviation has brought with it a global connection unprecedented in history. Global aviation today plays an integral role in global economic growth and benefits to our society. Aviation promotes commerce, provides 65.5 million jobs and $2.7 trillion to the Global Domestic Product (GDP). Aviation has effectively made the world a smaller place by connecting people in short time frames as a global community.

2.1. Today's world fleet

Today's global aviation fleet is composed of 25,830 airplanes with a mix of airplane capability for both passengers and cargo, with various flight ranges that include those below. Table 2 shows the Commercial Airplanes Forecast from 2019 to 2038.[k]

(1) Regional jets, short to medium range, up to 100 seats, e.g., Embraer E-190.
(2) Narrow body jets with a range capability of 1,500 to 3,800 miles, with up to 230 seats, e.g., Boeing 737.
(3) Wide body jets with a range capability of 7,525 to 8,690 miles, with 250 to 414 seats, e.g., Boeing 777.
(4) Freighters that are derivatives of the wide body jets but dedicated to hauling cargo rather than passengers.

2.2. Future fleet growth

Airlines will need ~44,000 new airplanes valued at $6.8 trillion over the next 20 years to meet the projected growth forecast for future world demand shown in Fig. 5. Long-term passenger traffic growth is estimated to be 3.4% over the same period. It is anticipated that by 2038, the world-wide fleet of airplanes will double. As the market grows, manufacturers will continue to introduce new products or derivatives that are more

---

[k]Boeing, "Boeing commercial market outlook, 2019–2038", The Boeing Company. https://www.boeing.com/resources/boeingdotcom/commercial/market/commercial-market-outlook/assets/downloads/cmo-sept-2019-report-final.pdf. For future market outlook reports download at: https://www.boeing.com/commercial/market/commercial-market-outlook/.

Table 2. Boeing commercial market outlook 2019–2038: Airplanes forecast on a page.

	Asia-Pacific	Asia-Pacific Detail					North America	Europe	Middle East	Latin America	Russia & Central Asia	Africa	World
		China	Southeast Asia	South Asia	Northeast Asia	Oceania							
Economic Growth Rate (GDP)	3.9%	4.7%	4.4%	5.8%	1.2%	2.4%	1.9%	1.6%	3.2%	2.9%	2.0%	3.4%	2.7%
Airline Traffic Growth Rate (RPK)	5.5%	6.0%	5.9%	7.4%	1.9%	3.7%	3.2%	3.6%	5.1%	5.9%	3.3%	5.9%	4.6%
Airline Fleet Growth Rate	4.6%	4.5%	5.7%	7.3%	1.5%	2.6%	1.9%	2.9%	4.9%	3.9%	2.1%	4.0%	3.4%
Market Size													
Deliveries	17,390	8,090	4,500	2,560	1,420	820	9,130	8,990	3,130	2,960	1,280	1,160	44,040
Market value ($B)	2,830	1,300	710	365	315	140	1,155	1,370	725	395	160	175	6,810
Average value ($M)	160	160	160	140	220	170	130	150	230	130	130	150	150
Unit share	39%	18%	10%	6%	3%	2%	21%	20%	7%	7%	3%	3%	100%
Value share	42%	19%	10%	5%	5%	2%	17%	20%	11%	6%	2%	3%	100%
Deliveries													
Regional Jet	210	120	20	10	50	10	1,680	80	20	30	180	40	2,240
Single Aisle	13,030	5,960	3,650	2,180	680	560	6,140	7,260	1,620	2,640	910	820	32,420
Widebody	3,810	1,780	820	370	590	250	850	1,540	1,440	270	140	290	8,340
Freighter	340	230	10	<5	100	<5	460	110	50	20	50	10	1,040
Total	**17,390**	**8,090**	**4,500**	**2,560**	**1,420**	**820**	**9,130**	**8,990**	**3,130**	**2,960**	**1,280**	**1,160**	**44,040**

(*Continued*)

Table 2. (*Continued*)

	Asia-Pacific	China	Southeast Asia	South Asia	Northeast Asia	Oceania	North America	Europe	Middle East	Latin America	Russia & Central Asia	Africa	World
			Asia-Pacific Detail										
Market Value ($B)													
Regional Jet	10	5	<5	<5	5	<5	85	5	<5	<5	5	<5	105
Single Aisle	1,535	680	450	255	85	65	710	845	190	310	95	90	3,775
Widebody	1,180	550	255	110	190	75	240	485	520	80	40	85	2,630
Freighter	105	65	5	<5	35	<5	120	35	15	5	20	<5	300
Total	**2,830**	**1,300**	**710**	**365**	**315**	**140**	**1,155**	**1,370**	**725**	**395**	**160**	**175**	**6,810**
2018 Fleet													
Regional Jet	140	50	<5	10	50	30	1,840	260	50	90	200	130	2,710
Single Aisle	5,680	3,050	1,110	580	570	370	4,130	3,740	670	1,240	770	400	16,630
Widebody	1,710	590	390	100	510	120	670	950	750	160	130	150	4,520
Freighter	350	200	30	10	80	30	910	310	80	90	170	60	1,970
Total	**7,880**	**3,890**	**1,530**	**700**	**1,210**	**550**	**7,550**	**5,260**	**1,550**	**1,580**	**1,270**	**740**	**25,830**
2038 Fleet													
Regional Jet	260	130	30	10	70	20	1,680	80	30	30	330	90	2,500
Single Aisle	13,950	6,560	3,670	2,340	800	580	7,060	7,070	2,000	2,890	1,210	1,020	35,200
Widebody	4,080	1,870	880	470	590	270	990	1,720	1,860	340	200	370	9,560
Freighter	1,130	770	90	60	160	50	1,200	470	140	120	200	140	3,400
Total	**19,420**	**9,330**	**4,670**	**2,880**	**1,620**	**920**	**10,930**	**9,340**	**4,030**	**3,380**	**1,940**	**1,620**	**50,660**

Source: http://www.boeing.com/commercial/market/commercial-market-outlook.

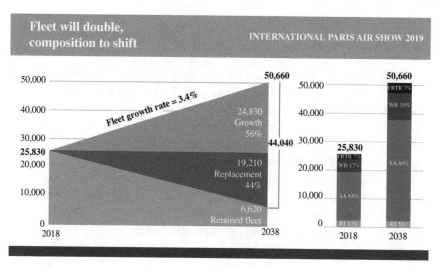

Fig. 5. Fleet Growth through to 2038.

Note: FRTR: Freighter, SA: single aisle, WB: widebody, RJ: Regional Jet.

Source: Boeing Aircraft.

energy efficient. Many factors weigh into business decisions for more fuel efficient airplanes; there are dependencies on new technology, business and airline strategies, and market drivers such as fuel price, environment and production rates.

To sustain healthy aviation economics, airlines want product advances at lower ownership and operational cost. The profits of today are largely due to technology investments from the prior decade, due to an airplane's long product lifecycle. It can take up to 10 years from concept to delivery of a new aircraft design. What we choose to invest in today will determine energy and fuel efficiency for the next decade.

Despite the challenges of new product launches, the introduction of new commercial jets and derivatives continue. Each new aircraft developed is typically 15–25% more efficient than the airplanes they replace. Over that time, research and development has resulted in significant advances in technology. The long product development cycle requires the need for technology readiness on the shelf. Aviation relies on a robust industry and collaboration between the government and academia. Significant technology advances have been achieved early and is readied for incorporation into new products.

3. Aircraft Energy Use

3.1. *Aircraft fuel efficiency drivers*

The fuel efficiency of a commercial transport aircraft, measured in terms of liters of fuel consumed per passenger per kilometer, is the primary measure used to evaluate aircraft performance. For a typical airline, fuel accounts for up to 40% of an airline's operating cost. Figure 6 illustrates the ICAO's total airline operating cost breakdown.[1]

The engines burn the fuel, but each of the aircraft subsystems (flight deck, avionics, environmental control systems, interior features, etc.) can be designed to improve fuel efficiency as well. Globally, the industry has adopted fuel efficiency certification standards (similar to noise and emissions) as a way to pressure aircraft manufacturers into developing technologies that improve efficiency and phase out older aircraft models that do not meet the industry standards, see Figs. 5 and 6. These standards are evaluated every three years and are adjusted to make sure fuel consumption goals are being met.

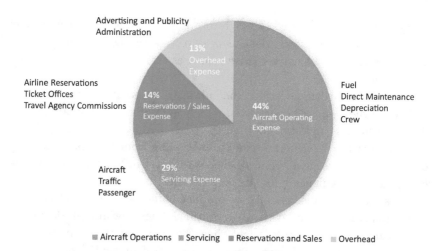

Fig. 6.   The ICAO's total airline operating cost breakdown.

---

[1]International Civil Aviation Organization, "Airline operating costs and productivity", International Civil Aviation Organization. https://www.icao.int/MID/Documents/ 2017/Aviation Data and Analysis Seminar/PPT3 - Airlines Operating costs and productivity.pdf.

An aircraft's fuel efficiency is driven by aerodynamic performance and propulsor and thermal efficiency while balancing adherence to regulatory certification and operational constraints, as well as international community noise and emissions standards (see the Appendices for details). The aircraft's flight deck, avionics, interior and payload system architectures are second-order contributors to meeting fuel efficiency goals.

Near term aerodynamic improvements for conventional tube and wing aircraft configurations are focused on maintaining laminar flow on wings, engine nacelles, empennages and winglets, minimizing/controlling separated flow regions through active and passive flow control technologies, and optimizing surface junctures such as the fuselage aft-body shape and inlet and exhaust devices.

Weight reduction is one of the key technology areas required to enable the challenging 2% per year fuel efficiency goal. The design, manufacturing and integration of new metals and new aeroelastically tailorable and/or multifunctional and adaptive composite materials are needed to enable the integrated, fuel efficient vehicles of the future.

3.1.1. *Aerodynamic performance*

Improving aerodynamic performance requires discovering ways to optimize the motion of air over the aircraft (see Fig. 7) by reducing the aerodynamic drag. The configuration of the aircraft, component design details, and the aircraft operational procedures all must be considered to improve aerodynamic performance. Given a specific mission profile, the aircraft can be designed with a wing/body/engine configuration to perform the mission with the best aerodynamic performance. Over the years, as commercial aircraft have evolved, advanced technologies have been developed to reduce the fuel burn (thus, improving performance) for commercial operations. A few of them are listed here:

Winglets: Aircraft wings producing lift form a large, powerful vortex at the wingtip and this vortex causes an aerodynamic lift-induced drag that increases fuel burn. Winglets were developed to reduce the size of that wingtip vortex and therefore reduce the amount of lift-induced drag.

Today's winglet on the Boeing 737 MAX, shown in Fig. 8, was designed with both an upper and lower winglet that optimizes the wingtip configuration and reduces fuel burn by up to 6%.

Fig. 7. Aircraft aerodynamics.

Fig. 8. Winglets.

Since 2000, aircraft retrofitted with winglet devices has saved over 80 million tons of CO_2 (see footnote c).

Natural Laminar Flow Technology: Air flowing over the aircraft scrubs the surface and produces aerodynamic viscous drag. At the leading edge of a wing, the flow at the surface starts out laminar (smooth) but quickly transitions into a turbulent flow due to pressure gradients and surface imperfections as illustrated in Fig. 9. A laminar flow produces less viscous drag than a turbulent flow. If the shape is designed correctly and the wing is built with a smooth surface, the transition to turbulent flow can be delayed and a natural laminar flow wing design that is more fuel efficient can be created.

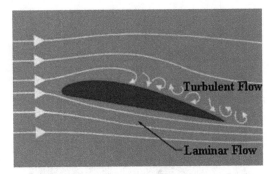

Fig. 9. Laminar flow transition to turbulent flow.

3.1.2. *Engine performance*

Aircraft engines are highly integrated into the aircraft design. Engine and aircraft manufacturers work closely together to design aircrafts that operate the required missions and provide value to their customers. Near team propulsion improvements for conventional tube and wing aircraft configurations are focused on increasing propulsor efficiency of the turbofan by reducing the Fan Pressure Ratio (FPR) to achieve improved fuel burn and reduced emissions and community noise levels. A lower FPR increases the fan diameter size to produce the same thrust. Improvements in the thermal efficiency of a turbofan engine are driven by increasing the overall pressure ratio (OPR). This highly efficient turbine engine core can turn a much larger fan that provides a large portion of the thrust. The challenge of today's aircraft designers is to find a way to fit these large fan diameter engines under the wing as shown in Fig. 10 of the new Boeing 777X aircraft (with GE9x engine) and a 134-inch diameter fan under its wing. Highly sophisticated design tools are used to design the wing and engine installation to optimize the structural engine mounts and to maximize fuel energy efficiency and reduce noise.

3.1.3. *Flight deck, avionics and systems*

The aircraft cockpit (sometimes called the flight deck) houses aviation electronics (see Fig. 11), which are the electronic systems that the aircraft pilot uses for flight controls, monitoring, communications, navigation, weather monitoring, and airspace/terrain situational awareness. New technologies use these electronic systems (also called avionics) to optimize the

Fig. 10.   Engines.

BOEING GRAPHIC

Pilots have applauded the flight deck of the Boeing 787 Dreamliner.

Fig. 11.   Aircraft interiors.

performance of the airplane when it is either in the air or on the ground. Today's aircrafts collect "big data" from the engines and all major onboard systems. This data is analyzed in real-time by using satellite connectivity. Relevant information is then used to inform pilots and ground operation staff of opportunities for improving operational efficiency. For example, today's connected airplane technology not only provides pilots with updated weather information along the route, it also provides optional routes that take advantage of favorable winds and weather, traffic conditions on approach and the ground at their destination, and aircraft health monitoring.

New technology is replacing pneumatic and hydraulic systems with electrically powered systems. The primary differentiating factor in the system architecture of the Boeing 787 is its emphasis on electrical systems.[4] One of its advantages is the greater efficiency gained in terms of reduced fuel burn; the less draw on engine air improves fuel efficiency. The Boeing 787's system architecture accounts for fuel savings of about 3%. The Boeing 787 also offers airlines operational efficiencies due to the advantages that electrical systems have over pneumatic systems in terms of weight and reduced lifecycle costs.

3.1.4. *Interiors and payloads*

An airplane's weight plays a key role in overall fuel efficiency — its interiors and payloads may not seem to have significant leverage to improve airplane fuel efficiency, but "lightening the load" is a proven and practical method to achieve the goal. Airlines are discovering that every gram counts. For example, Virgin Atlantic estimated that losing a pound (0.45 kg) in weight from every plane in its fleet would save 53,000 liters of fuel a year, adding up to tens of thousands of dollars in savings.[5]

In the last 10 years, the use of composites in aircraft interiors has not kept up with their use in aircraft structures.[m] Given the composites' unique combination of low flammability, low weight, high strength, durable cosmetics and energy efficiency with which they can be processed in the manufacturing environment, composite materials have a promising future in passenger aircraft interiors as shown in Fig. 12 at Boeing's South Carolina's Interiors Responsibility Center (IRC). The OEM and aftermarket interiors segments together represent a market for ~22.5 million lb. (~10.2 million kg) of composites this year.

On the Boeing 787, many of the fixtures are produced using a material called the Nomex honeycomb core,[n] which is made with a structure of aramid fiber paper coated in a resin and cured, creating a product that is both rigid and incredibly lightweight. Once the material is pressed into shape and refined, the necessary supporting structures and hardware are added, such as door hinges and handles.

---

[m]Composite World, "Composites in aircraft interiors, 2012–2022", *Composite World*, 9 April 2012. https://www.compositesworld.com/articles/composites-in-aircraft-interiors-2012-2022.

[n]"A look behind the scenes at Boeing's IRC". https://runwaygirlnetwork.com/2018/04/03/a-look-behind-the-scenes-at-boeings-interiors-responsibility-center/.

Fig. 12. A sample fixture made of Nomex honeycomb core being produced at the Boeing's South Carolina IRC.

In addition to the use of composites in airframe primary structures, the interior composite market is twice as big in terms of its consumption of composite materials and manufacturing services. By 2022, this is expected to grow beyond 32 million lb. (14.5 million kg) of components annually — a figure that is about 50% greater than the total demand for composite airframe structures in 2012. Energy use for production and disassembly as well as recycling are future opportunities for improvement.

Water use on commercial aircrafts is an area where new technology can be used to reduce the energy consumed on each flight.[6] During flights passengers use drinking water for different purposes. If more water is carried on the aircraft than is used on the flight, additional energy must be used to carry that water. Advanced technology can measure the amount of water used on each flight and provide guidance for the planning of future flights that will reduce the amount of water onboard upon arrival. An airplane wash is required on a regular basis to maintain aerodynamic efficiency — most often using water. One technology being tested is a new paint technology that resists dirt and grime, and therefore reduces the amount of water to keep an aircraft clean.

3.2. *Aircraft production drivers*

Airplanes are produced by a large-scale integration of vehicle components with millions of parts. For instance, the Boeing 777 has over three million parts that are brought together for final assembly in Boeing's manufacturing facility in Everett, Washington. The airplane's assemblies rely on raw materials, components, systems, subsystems and subassemblies, and are manufactured either on site or at different manufacturing facilities around the world by a global supply chain.

A product's energy production footprint would need to consider the global totality of the materials, manufacturing and aircraft integration activities required to deliver the final product. Manufacturers' manage energy consumption through extensive energy conservation and promote energy efficiency technology and infrastructure improvements. New construction at many sites strive for the Leadership in Energy and Environmental Design (LEED) certification, which will help focus design of new construction towards high performance, building towards energy goals over the lifetime of the facility.

3.2.1. *Materials*

Aircraft manufacturers rely on suppliers for many essential materials, parts and subassemblies. Some of the largest volume is found in the primary structural materials of aluminum, titanium (e.g., sheet, plate, forgings and extrusions) and composite materials (including carbon and boron fibers). Newer large aircrafts rely heavily on composite materials due to their structural efficiency to weight ratio providing lighter weight aircraft, and hence, significantly better fuel efficiency. As shown in Fig. 13, Boeing's latest large aircraft introduced into service in 2010, the Boeing 787 consists 50% of advanced composites.[o] The Airbus A350 XWB will be 53% composite material.

Aside from the primary structure, a myriad of materials are found on an aircraft that has a history of performance and is qualified for a specific purpose. It is often a challenge to move on to new materials due to the necessary equivalent performance and qualification needed to ascertain equivalent safety.

Meanwhile, ongoing research has been focused on moving to more sustainable materials with an overall reduced life cycle environmental footprint. Life cycle analysis tools are becoming more available for designers to consider different material solutions that may offer a reduced energy

---

[o]AERO, "Boeing 787: from the ground up", *AERO*, 2006. https://www.boeing.com/commercial/aeromagazine/articles/qtr_4_06/index.html.

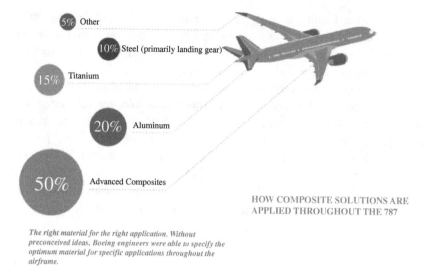

The right material for the right application. Without
preconceived ideas, Boeing engineers were able to specify the
optimum material for specific applications throughout the
airframe.

Fig. 13.   Composites in the airframe and primary structure.

footprint over the life of material usage. For instance, certain materials
might point a designer towards using a recycled material that does not
necessarily require high-energy usage associated with raw material
extraction.

Design considerations for recycling and reclamation can also have an
impact on material disposition. An ability to recover material from the
airplane for reuse must be designed-in to ensure it can be done at a reasonable labor cost. Technology research in energy efficient materials recovery continues in the reclamation industry. Additionally, the preference for
non-hazardous materials is a benefit from a handling/worker safety perspective, as well as from an environmental disposition standpoint, which
can have energy implications.

3.2.2. Manufacturing

Manufacturers and their suppliers must adhere to stringent standards and
practices, and certify that their production systems meet the requirements of manufacturer quality management systems approved by the
FAA. Each manufacturing facility depends on the availability of local

energy sources, such as electricity. Sources of energy rely on regional utilities, and manufacturers work locally to ensure energy availability and affordability.

Some processes can require significant energy at manufacturing facilities, such as the autoclave, a large oven-like compartment to cure large composite structures, which requires high temperatures for protracted duration for large volume. Also, paint facilities are large buildings that house an airplane during the final painting of the airplane livery. Aircraft coatings require a specific temperature range to ensure that paint and coatings harden at appropriate temperatures, so corrosion inhibiting characteristics can be maintained.

There are additional operations in the factory that require energy. Efforts are made during the design of aircraft to minimize energy requirements during the manufacturing process to reduce facility energy requirements and to help manage overall production costs. Technology development aimed at addressing high-energy requirements during production can also reduce the overall energy footprint of manufacturing.

In addition, the proximity of suppliers and the energy impact of moving parts long distances are part of the product's lifecycle environmental footprint.

3.2.3. *Aircraft integration*

Airplane performance and manufacturing capability are considered early and concurrently to ensure that optimum designs aimed at fuel efficiency are also buildable in the factory. Some design decisions are constrained by the ability to build a configuration at a reasonable cost and with an energy footprint with projected production rates to meet the business case that can bring a product to market.

As found in many industries, the manufacturing energy footprint is often dwarfed by "product in use" energy. For instance, with clothing, the energy footprint of washing clothes over a lifetime is much larger than the energy required to make the original clothing. So design and manufacturing decisions must consider the entire lifecycle of product integration. An airplane will typically fly daily for 20 or more years, so expending some additional manufacturing energy to build a more weight efficient airplane that will fly the entire 20 years of its service life is an overall life cycle energy benefit.

3.3. *Aircraft in-service usage: Operational efficiency drivers*

After an airplane is delivered, the operator assumes responsibility for the aircraft operation and the associated energy required to fly and maintain the airplane. Fuel efficiency becomes a priority focus for the operator since fuel costs are a significant cost for airlines (see Fig. 7).

Operational efficiency energy can include the energy footprint of flight operations, ground operations and maintenance operations. Time on the ground also requires energy for servicing, boarding and deplaning at the airport. This section is focused on the gate-to-gate energy implications, when the aircraft pushes back from the gate until it reaches and parks at its destination gate. The airport energy footprint is not addressed in this section but plays an important role as regions strive to serve more passengers and cargo. Airports, like manufacturers, practice energy conservation, with many improvement and renewable energy projects to seek energy management gains that align with economic and social implications for their region.

3.3.1. *Utilizing airplane capability*

Aircraft operators and manufacturers work together to fully utilize capabilities of the airplane to move passengers and goods most efficiently. The airline ground systems, in coordination with flight deck systems, are used to meet these goals. The airplane's flight deck capability includes the flight management computer and avionics that enables flight crews to seek optimal operations. In the last decades, many navigation aids and the mainstreaming of global positioning satellite (GPS) are also providing airplane capability to aid in-flight operations and flight deck decisions that can improve overall flight fuel efficiency.

3.3.2. *Procedures and gate to gate efficiency*

Routing procedures such as direct routes and Area Navigation (RNAV) routes allow shorter, more efficient paths to be flown.[p] Optimized arrival and departure procedures can simultaneously reduce community noise,

---

[p]Federal Aviation Adminstation, '"NASA and FAA celebrate transfer of new technology", Federal Aviation Adminstation, 28 September 2018. https://www.faa.gov/news/updates/?newsId=91707.

Box A. Efficient aircraft operations

The aircraft operations consist of a broad range of activities related to all stages of air travel (i.e., on the ground before passengers and cargo are loaded, taxi-ing to and from the runway, departure, climb, cruise, descent and landing).

Optimization of operational procedures has the potential to reduce emissions through the minimization of the amount of fuel used in each flight by:

- Taxiing and flying the most fuel-efficient route;
- Operating at the most economical altitude and speed;
- Maximizing the aircraft's load factor;
- Loading the minimum fuel to safely complete the flight;
- Minimizing the number of non-revenue flights; and
- Maintaining clean and efficient airframes and engines.

Source: Federal Aviation Administration, "NASA and FAA celebrate transfer of new technology", *Federal Aviation Adminstation*, 28 September 2018. https://www.faa.gov/news/updates/?newsId=91707.

emissions, fuel, and even time for both air and ground crews. Figure 14 illustrates an example, a technology called the Flight Deck Interval Management (FIM), originally developed by NASA and in development by the FAA and airlines to increase capacity, reduce delays, fuel burn and emissions. Boeing, Honeywell and United Airlines participated in the FIM development and flight test that involved two aircrafts from Honeywell's flight test fleet as well as a United Boeing 737.

In the near future, digital communications will be used to better coordinate air and ground operations to enable even more efficient use of airplanes. In addition, improved airplane landing capabilities improve efficiency by not holding or diverting airplanes in poor weather.

The combined new air and ground capabilities allow more fuel and noise efficient routings of airplanes, and more efficient operations of airline fleets. Getting an airplane quickly to a gate that is already in use benefits neither the passengers nor efficiency. Shared information among the airline operators, airspace controllers and the airplanes allows improvements in overall operational efficiency. This coordination takes advantage of

Fig. 14.   NASA/FAA Flight Deck Interval Management (FIM).

airplane capabilities already deployed to improve noise, fuel and time efficiency.

Improved routing efficiency, gate coordination, reduced holding times, and ground equipment improvements can all lead to additional efficiency benefits.

3.4. *Improved efficiency by generating meteorological data from aircraft systems*

In recent years there have been advancements in airplane connectivity, information processing, displays for visualization, and improved atmospheric modeling techniques. These technologies are creating benefits for safety enhancements, more fuel-efficient operations, on-time performance, reduced maintenance, and increased ride comfort for passengers. The typical comprehensive meteorological observations include winds, temperature, turbulence, humidity, icing and other cloud properties as shown in Fig. 15. Airlines typically share their observations with each other as well as the worldwide national weather services for assimilation into forecast models and datasets. The end use of this weather and turbulence information creates improved decision support tools for airline operations

Fig. 15.   Meteorological data.

(e.g., dispatchers) as well as for pilots on their portable and installed displays.

3.5. *Aircraft end of service*

To minimize the energy consumed on an aircraft at its end of service, we need to rethink our approach and develop new technology needed to reduce, reuse and recycle airframes. This will also extract the full residual value of the aircraft at the end of its service. Aircraft are designed to be in service for more than 20 years. By 2050, older airplanes will be retiring at a rate of 1,500 aircraft per year (compared with about 500 aircraft per year today).[q] Just like every other consumer product, there is a life cycle for every aircraft. This includes the aircraft's design, built, and in-service and end of service years. A tremendous amount of intellectual resources are focused on designing, building and operating an efficient aircraft. Over the last 10 years, the industry has started to focus on the End of Service portion of the aircraft's life cycle, and organizations have been developed to address it.[7] The Aircraft Fleet Recycling Association (AFRA) is the leading global organization for developing the safe and sustainable management of end-of-life aircraft and its components. The focus of these efforts is to recycle and/or reuse as much of the materials in the End of Service aircraft and therefore reduce the amount of materials sent to landfills.

---

[q]ALTON Aviation Consultancy, "Global Commercial Fleet and MRO Dynamics and Trends", 31 Oct–2 Nov 2017 Aviation Week MRO Asia-Pacific, Singapore https:// altonaviation.com/alton_insights/global-commercial-fleet-and-mro-dynamics-trends-2/.

Fig. 16.   Boeing's 2015 ecoDemonstrator 757 being recycled.

3.5.1.   *Recycling and repurposing of materials*

The vast majority of a commercial aircraft, estimated at 85–90% of its weight, is recyclable today. The remaining 10–15% portions of the aircraft, which tend to be interior components embedded with flame retardants as required by the FAA, are sent to landfills. Solutions are quickly being developed across the industry, with more promising providers and technologies each year. Figure 16 shows the Boeing 2015 ecoDemonstrator 757, which was recycled in collaboration with the AFRA and Aircraft Demolition with 93% recycled or reused by weight. Benefits of recycling include:

- An airplane's aluminium is 90% more energy efficient than raw production.
- Recycling and reuse help lower exposure to supply vulnerability of rare earth metals, titanium and other core materials that are derived from growing competitors or conflict-prone regions in developing countries.[8]

Opportunities to further improve aircraft decommissioning include increased rates among providers using best practices, such as being certified to the AFRA standard.[r]

---

[r]Aircraft Fleet Recycling Association, "Accreditation", Aircraft Fleet Recycling Association. https://afraassociation.org/accreditation/.

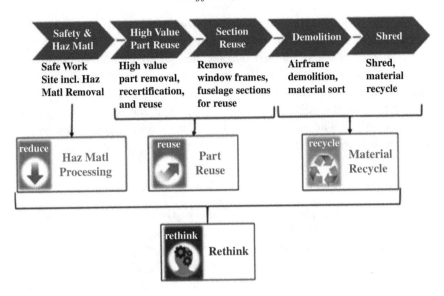

Fig. 17.   Standard commercial aircraft end of service process.

The AFRA standard requires responsible practices in processes from tagging and documentation to hazardous material disposal. Airlines and aircraft owners can request for the AFRA certification from their vendors to ensure that their aircrafts are retired with adherence to strict environment and safety protocols.

Figure 17 illustrates the standard End of Service process for commercial aircrafts. The disassembly process for End of Service aircraft should maximize the value of recovered materials, mitigate energy, and reduce the disposal risk of residual hazardous materials.

When it comes to recycling aircraft materials, the ideal situation would be to take the metal from an End of Service aircraft and return it to the supply chain as raw metal stock. Today, there are significant challenges in doing so. Because each component of the aircraft is made up of many different alloys, separating them into individual alloys is a technology challenge. New technologies and processes are enabling the use of recycled and scrap materials back into the factories from which they came, or to other industries. Today, in large recycling facilities, the use of new technology sensors to identify metals through infra-red scanning and X-rays has become popular. Three common categories of metal sensing processes are: biotechnology (biology based metal separation), hydrometallurgy (chemical metal separation), and pyrometallurgy (high temperature metal separation).

About 1,500 components can be removed from an aircraft and sold on the second-hand market. In total, they account for 20–30% of the weight of the aircraft. Clearly marked parts in areas that do not degrade would assist in the amount of parts available for reuse after disassembly. Companies that specialize in Aviation Aftermarket Sales and Engineering are certified to the AS9120 standard to procure and sell parts, materials and assemblies. The standard focuses on traceability, control of records and airworthiness certificates, among others, for a safe return to service.

3.5.2. *Informing design for environment*

Other efforts are in place to try to improve the ability to recycle and/or repurpose materials. This includes the early design phase to provide aircraft designers with information to help select materials with recycling in mind. Selecting common or similar materials in major aircraft components (e.g., fuselage skins) would make the recycling of aircraft fuselage into high value materials, achievable and result in overall energy reduction.

4. The Role of US Federal Agencies: Leveraging Public-Private Partnerships

The United States' Federal Agencies have accelerated the maturation and adoption of fuel-efficient technologies to the private, commercial subsonic fleet (passengers and cargo) and military market segments using public-private partnerships. These partnerships may lead to fuel efficiency improvement on the order of 50–70% below current levels.[s]

4.1. *National Aeronautics and Space Administration (NASA)*

During the fiscal years from 2010 to 2015 (FY10–15), NASA's Advanced Air Vehicles Program (AAVP) and Integrated Aviation Systems Research Program (IASP) conducted research on advanced vehicle concepts and technologies that will improve fuel efficiency and reduce emissions and community noise.

---

[s]International Civil Aviation Organization, "United States aviation greenhouse gas emissions reduction plan", International Civil Aviation Organization, June 2015. https://www.icao.int/environmental-protection/Lists/ActionPlan/Attachments/30/UnitedStates_Action_Plan-2015.pdf.

Box B.   NASA's Environmentally Responsible Aviation (ERA) Project

NASA's Environmentally Responsible Aviation Project's portfolio included:

- New embedded nozzles that blow air over the surface of the airplane's vertical tail fin; this enables planes to be constructed with smaller tails to reduce weight and drag.
- New process for stitching together composite materials to create more damage resistant structures, so that planes can be shaped differently to reduce overall weight and enable revolutionary vehicles like the Hybrid Wing Body (HWB).
- New shape-changing wings that improve the design of airplane flaps that will enable them to extend more efficiently.
- Improved parts of a turbine engine that can improve aerodynamic efficiency and improve fuel burn.
- Improved the design for a small core engine fuel-flexible combustor, the chamber where fuel is burned, to reduce the amount of nitrogen oxide produced to cut overall pollution. Tested conventional and alternative fuel testing of advanced combustor configurations and assessed fuel injection performance, operability, durability and thermal growth management performance.
- New advanced fan design to improve propulsion, reduce fuel burn and reduce noise in jet engines.
- New designs to reduce noise from wing flaps and landing gear.
- The feasibility and the efficacy of the HWB advanced vehicle concept in which the wings seamlessly join the fuselage, and the engines are mounted on top of the rear of the airplane to shield it from engine noise.

Source: https://www.nasa.gov/press-release/nasa-research-could-save-commercial-airlines-billions-in-new-era-of-aviation/.

4.1.1.   *The Environmentally Responsible Aviation Project*

The National Aeronautics and Space Administration's (NASA's) Integrated Aviation Systems Program included the Environmentally Responsible Aviation (ERA) project, a public-private partnership focused on subsonic vehicles.[9] The project conducted research, technology development and demonstrations for advanced airframes including both traditional (i.e., tube and wing) and unconventional (i.e., HWB) prototypes. The ERA's portfolio encompassed the full life cycle of technology development from lab specimens to large scale ground tests (i.e., wind tunnel, high pressure sector tests, annular engine test cells, and structural iron wings) and/or flight tests to achieve Technology Readiness Level [TRL]) 5/6.[t] (See TRL Levels in Appendix A.) Over the course of the 6-year-old project, NASA invested more than $420 million, with a further $230 million of in-kind resources contributed by industry and federal lab partners. Key performance goals achieved include the reduction of aircraft drag by 8%, aircraft weight by 10%, and specific engine fuel consumption by 15%. The ERA Technology Portfolio demonstrated simultaneously a 47% fuel burn reduction, 79% LTO NO_x emissions reduction to CAEP 6 guidelines and 40.7 EPnDB noise reduction to Stage 4 guidelines on a 2025 advanced vehicle (Hybrid Wing Body [HWB] with a Geared Turbo Fan), which was compared against a Best in Class 2005 B777-200 with a GE direct-drive engine configuration.[u]

4.1.2. *Advanced Air Transport Technology Project*

Current NASA research builds upon ERA research with its Advanced Air Transport Technology (AATT) Project with a focus on vehicle efficiency. Formerly known as the Subsonic Fixed Wing (SFW) Project, this research is focused on achieving the Green Aviation goals shown in Fig. 18 for the "Aircraft Entering Service in the 2025–2030" timeframe. The AATT Portfolio includes research in the following areas:

- Aerodynamic and structures technologies that optimize the wing aspect ratio by 50 to 100% and reduce weight and drag while retaining aeroelastic integrity and flight control authority.
- Technologies that could make hybrid-electric propelled aircraft possible.
- Distortion-tolerant inlet/fan and leverage boundary-layer ingestion concepts to achieve a vehicle-level net system benefit.
- Advance small core engines.

---

[t]National Aeronautics and Space Administration, "Technology readiness level definitions", National Aeronautics and Space Administration. https://www.nasa.gov/pdf/458490main_TRL_Definitions.pdf.
[u]National Aeronautics and Space Administration, "Environmentally responsible aviation — end of an era, part 2", video, National Aeronautics and Space Administration. https://www.nasa.gov/mediacast/nasa-x-environmentally-responsible-aviation-end-of-an-era-pt2.

TECHNOLOGY BENEFITS	TECHNOLOGY GENERATIONS (Technology Readiness Level = 5-6)		
	Near Term 2015-2025	Mid Term 2025-2035	Far Term beyond 2035
Noise (cum below Stage 4)	22 – 32 dB	32 – 42 dB	42 – 52 dB
LTO NO_x Emissions (below CAEP 6)	70 – 75%	80%	>80%
Cruise NO_x Emissions (rel. to 2005 best in class)	65 – 70%	80%	>80%
Aircraft Fuel/Energy Consumption (rel. to 2005 best in class)	40 – 50%	50 – 60%	60 – 80%

Fig. 18. NASA's Mid-Term (2025–2035) environmental reduction goals for its Advanced Air Transport Technology project.

Source: National Aeronautics and Space Administration, "NASA aeronautics: strategic implementation plan–2017", update. NP-2017-01-2352-HQ, National Aeronautics and Space Administration. https://www.nasa.gov/sites/default/files/atoms/files/sip-2017-03-23-17-high.pdf.

- Foundational research on combustion physics to gain a better understanding on particulate matter and NO_x reduction in the cruise phase. Technologies include active combustion control, and lean direct injection (LDI) using conventional and alternative fuels.
- Non-traditional aircraft wing and body shapes.
- More widespread use of composite materials in building lighter weight aircraft with a focus on high-rate manufacturing.

4.1.3. *Unconventional commercial subsonic aircraft concepts*

Through NASA-sponsored research in the SFW/AATT and ERA Projects, sponsored advanced vehicle concept trade studies were conducted that leveraged the technologies identified in Appendix A. The airframe, propulsion and operations technologies were integrated into advanced commercial subsonic aircrafts, both traditional tube and wing configurations as well as unconventional configurations as shown in Fig. 19. These studies evaluated these advanced configurations against NASA's far-term N+3 (¿2035) Subsonic Transport Goals (Fig. 4) for energy efficiency, emissions reduction and community noise reduction.

In the top left corner of Fig. 19 is a truss-braced wing configuration that incorporates advanced airframe technologies including laminar flow, high aspect ratio wing design, lightweight wing-strut structure and an

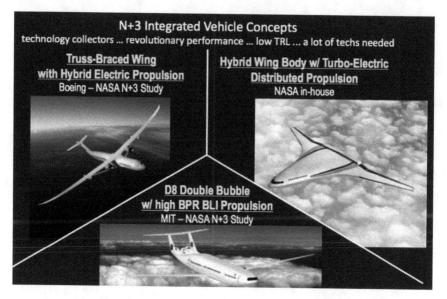

Fig. 19.   Integrated vehicle concepts (N + 3: 2035 + entry into service).
Source: https://www.hq.nasa.gov/office/aero/pdf/wahls_2_green_aviation_summit.pdf.

advanced propulsion system that leverages a hybrid-electric propulsion system. This vehicle concept is known as the Boeing Sustainable Ultra Green Aircraft Research (SUGAR) Volt and was produced by a Boeing-led team. In the top right corner is a variation of NASA's HWB configuration that incorporates a turbo-electric, distributed propulsion system. The HWB configuration by design optimizes propulsion/airframe integration to shield the aircraft from engine noise. The bottom image is known as the D8 Double Bubble configuration that incorporates high-aspect ratio wing design with a high by-pass ratio and a boundary layer ingesting propulsion system. Additional details on these N+3 Vehicles studies and others can be found in Refs. 10 and 11. A summary of potential airframe and propulsion technologies that are in various stages of research, development and product insertion are shown in Appendix B.

4.1.4.   *Turboelectric and hybrid electric aircraft*

With the emergence of an urban air mobility transport sector forecasted to begin in the early 2030s (as illustrated in Fig. 20), the aviation industry vehicle and engine OEMs — both domestic and international — are investigating electrified aircraft propulsion technologies and electrified

Fig. 20.   Urban air mobility landscape.

Source: https://www.nasa.gov/sites/default/files/thumbnails/image/rvlt_urbanairmobility.jpg.

vertical take-off and landing (eVTOL) aircraft concepts to meet market needs as well as target reductions in fuel burn, emissions and community noise.[12,13] A NASA-funded market study on the potential of urban air mobility (UAM) use cases identified as the last-mile delivery, air metro and air taxi passenger operations as the early adopters of electrified aircraft concepts.[v] There are many technological as well as regulatory challenges that need to be addressed to establish the market segment including: overcoming electric drive component weight and efficiency penalties, enabling autonomous flight controls development, establishing vehicle certification and regulatory standards (community noise) and integration of a diverse set of aircraft from small Unmanned Aerial Systems (UAS) (<55 lbs to 50+ passenger vehicles) into the next generation air-traffic controlled airspace.

---

[v]National Aeronautics and Space Administration, "Urban Air Mobility (UAM) market study", November 2018. National Aeronautics and Space Administration. https://www.nasa.gov/sites/default/files/atoms/files/uam-market-study-executive-summary-v2.pdf.

4.2. *Federal Aviation Administration*

The Federal Aviation Administration (FAA) supports an ambitious "The Next Generation Air Transportation System, [which] is the FAA-led modernization of America's air transportation system to make flying even safer, more efficient, and more predictable".[w] The Continuous Lower Energy, Emissions and Noise (CLEEN) Program is a key element.

Developed in 2010, the CLEEN is the FAA's environmental plan to accelerate development of technologies that will reduce noise, emissions, fuel burn and advance sustainable alternative jet fuels while allowing for sustained aviation growth. CLEEN technologies can be used in retrofits to current and future aircrafts.

The CLEEN Program is structured as a 50-50 cost-share public-private partnership. These industry-led partnerships implement ground and/or flight test demonstrations to accelerate maturation of certifiable aircraft and engine technologies. Figure 21 provides a summary of the CLEEN Phases I and II Program scope and environmental targets for fuel burn, noise and emissions.

The first CLEEN [Phase] I Program began in 2010 with five industry partners. Over a 5-year period, the FAA invested a total of $125 million. Coupled with partner funding, the total investment value exceeded $250 million. The CLEEN I Program had a 33% fuel burn reduction target

	Phase I	Phase II
Time Frame	2010–2015	2016–2020
FAA Budget	~$125M	~$100M
Noise Reduction Goal	25 dB cumulative noise reduction cumulative to Stage 5	
Fuel Burn Goal	33% reduction	40% reduction
NO_x Emissions Reduction Goal	60% landing/take-off NO_x emissions	75% landing/take-off NO_x emissions (−70% re: CAEP/8)
Entry into Service	2018	2026

Fig. 21.   CLEEN Program Phases I's and II's environmental goals.

Source: https://www.faa.gov/about/office_org/headquarters_offices/apl/research/aircraft_technology/cleen/.

---

[w]Federal Aviation Administration, "Modernization of US aerospace", Federal Aviation Administration. https://www.faa.gov/nextgen/.

Box C. Technology Portfolio accelerated by CLEEN Phase I Program.

- An adaptive trailing-edge treatment for aircraft wings that improves aerodynamic efficiency as well as reduces aircraft noise on approach.
- A ceramic matrix composite (CMC) acoustic nozzle at the engine exhaust that reduces component weight and maintains higher temperature operations that results in less fuel consumption and a quieter engine.
- Advanced engine combustor concepts that enable significant Landing and Takeoff Nitrogen Oxides (LTO NO_x) reduction.
- Advanced open rotor engine that can deliver fuel consumption savings due to the lower fan pressure ratio.
- A more efficient Flight Management Systems (FMS) that improves engine monitoring and performance.
- New seal and coatings technologies that enable an engine to operate at higher temperatures in the compressor and turbine, which result in improved thermal efficiency.
- An advanced ultra-high bypass (UHB) geared turbofan (GTF) engine that can deliver a more efficient engine that results in less fuel consumption and a quieter engine.
- Turbine section thermal efficiency improvements including dual-wall turbine airfoils and a CMC Blade Track, which results in increased operating temperature capability and reduced engine cooling requirements while gaining significant weight savings and less fuel consumption.

Source: https://www.faa.gov/about/office˙org/headquarters˙offices/apl/ research/aircraft˙technology/cleen/media/CLEENI˙CLEENII˙Projects.pdf.

---

relative to the best in class 2005 and a 60% LTO NO_x emissions reduction target relative to the CAEP/6 standard.

In 2015, the CLEEN Phase II Program was established with eight partners over a 5-year period. With cost sharing, the total federal investment is expected to be $100 million.[x] The CLEEN Phase II Program has a 40% fuel burn reduction target relative to the best in class 2005 and a 75% LTO NO_x emissions reduction relative to the CAEP/6 standard (as shown in Fig. 21). The CLEEN II technologies (as shown in Fig. 22)

---

[x]Federal Aviation Administration, "Continuous Lower Energy, Emissions, and Noise (CLEEN)", Federal Aviation Administration. https://www.faa.gov/about/office_org/ headquarters_offices/apl/research/aircraft_technology/cleen/.

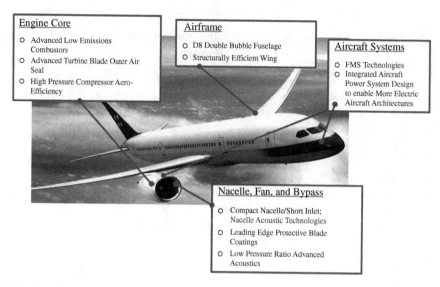

Fig. 22.   CLEEN II technologies.

Source: https://www.faa.gov/about/office_org/headquarters_offices/apl/research/air-craft_technology/cleen/media/CLE.ENI_CLEENII_Projects.pdf.

support the FAA's goal of achieving a net reduction in climate impact from the aviation industry.[y]

The follow-on CLEEN Phase III Program is focused on maturing previously conceived noise, emissions and fuel burn reduction technologies for civil subsonic and supersonic aircraft from TRLs of 3–5 to TRLs of 6–7 to enable the industry to expedite the introduction of these technologies into current and future aircraft and engines. New for CLEEN Phase III, as shown in Fig. 23, are civil supersonic transport noise and emissions goals for certifiable aircraft technology that reduce landing and takeoff cycle (LTO) noise levels and/or nitrogen oxide emissions for civil supersonic airplanes.

Subsonic transport goals include certifiable aircraft technologies that:

- Reduces noise levels by 25 dB cumulative, for civil subsonic airplanes and/or reduces community noise exposure, or
- Improves aircraft fuel efficiency by 20% relative to the ICAO's new type fuel efficiency standard for civil subsonic airplanes adopted in 2016,

---

[y]Federal Aviation Administration, "CLEEN I and II projects", Federal Aviation Administration, 2 February 2018. https://www.faa.gov/about/office_org/headquarters_offices/apl/research/aircraft_technology/cleen/media/CLEENI_CLEENII_Projects.pdf.

	Phase III*	
Time Frame	2021–2025	
Entry into Service	2031	
FAA Budget	TBD	
Vehicle Type	Subsonic	Supersonic
Noise Goal	25 dB cumulative noise reduction cumulative to Stage 5 and/or reduces community noise exposure	Reduction during landing and takeoff cycle (LTO)
Fuel Burn Goal	–20% re: CAEP/10 Std	–
NO_x Goal	–70% re: CAEP/8 Std (LTO)	Reduction in absolute NO_x emissions
Particulate Matter Goal	Reduction rel: CAEP/11 Std (LTO)	–
* The information for the third phase of the CLEEN Program is notional as the FAA is in the process of developing the final solicitation.		

Fig. 23.   CLEEN Phase III Program's environmental goals.

Source: https://anesymposium.aqrc.ucdavis.edu/sites/g/files/dgvnsk3916/files/inline-files/20200303%20UC%20Davis%20ANE%20Symposium%20-%20CLEEN%20Overview%20%28Dorbian%29.pdf.

- Reduces LTO cycle nitrogen oxide emissions for civil subsonic airplanes by 70% over the ICAO's standard adopted in 2010 (CAEP/8), and/or reduces absolute NO_x production over the aircraft's mission, and
- Reduces particulate matter relative to the CAEP/11 Standard (LTO).

CLEEN III also aims to advance the development and introduction of hydrocarbon jet fuels for aviation that could enable improvements in fuel efficiency and reductions in emissions. This includes fuel blends. The CLEEN Program is interested in fuels that are drop-in compatible with

---

Box D.   Perspectives on the US Air Force's Energy Use

- The Federal Government uses 1% of all US Energy consumption.
- The DOD uses 80% of all Federal energy use.
- The Air Force uses 48% of the DOD's energy use.
- Aviation accounts for 80% of the Air Force's energy use.

Source: United States Aviation Greenhouse Emissions Reduction Plan, submitted to ICAO, June 2015. https://www.icao.int/environmental-protection/Lists/ActionPlan/Attachments/30/UnitedStates_Action_Plan-2015.pdf.

---

the existing pipeline and airport-fueling infrastructure, but have changes in their composition that could help an aircraft meet these CLEEN Program's goals.

4.3. *Department of Defense*

The US Air Force has four energy priorities: improve resiliency, reduce demand, assure supply and foster an energy aware culture. The Air Force consumes roughly half of the fuel consumed by the Federal Government (see Box D). The Air Force Research Laboratory at the Wright Patterson Air Force Base focuses its technology development and research efforts in three primary areas: aerodynamic improvements, weight reduction and increased engine efficiency. Highlights include:

- The Surfing Aircraft Vortices for Energy ($AVE) formation flight technology that harvests energy from the lead aircraft's wing tip vortices, which has enabled 8–10% in fuel savings for the trailing aircraft.
- Weight reduction has been achieved through precise fueling, lighter weight equipment and removal of unnecessary or reductant equipment.
- The Adaptive Engine Technology Development (AEDT) Program will enable a 25% fuel consumption reduction in the Air Force's F-35 fleet upon the completion of its engine upgrade in 2023. This effort has a potential saving of over 1 billion gallons of fuel by 2040.

The Department of Defense's (DOD) AETD Program has a 25% fuel efficiency improvement goal for all military engines by 2020. It is anticipated that advances in military engine fuel efficiency technologies will transition to the commercial subsonic fleet in the mid 2020–2030s.

4.4. *Environmental Protection Agency*

The Environmental Protection Agency (EPA), established in 1970, is tasked with protecting human health and the environment, as directed by Congressional legislation.[14] The agency plays a central role in developing environmental regulations across all sectors. The EPA establishes regulations through "notice and comment" rule-making under which it first issues as draft proposals, receives public comments and then issues final rules. Domestically, the EPA is currently assessing whether the appropriate regulations for carbon dioxide standards in US aircrafts are as stringent as the ICAO's carbon dioxide international standards.

5. Alternate Fuels (Biofuels)

In 2019, The Commercial Aviation Alternative Fuels Initiative (CAAFI) celebrated its 10th anniversary in working towards sustainable jet fuels. The CAAFI coalition was founded by the FAA in 2006 to promote the development and deployment of alternative jet fuels that reduce lifecycle carbon emissions and local pollution around airports.

Alternative fuels such as aviation sustainable biofuels have been approved as a drop-in fuel for use on commercial aircraft at up to a 50/50 blend by the international standards organization, American Society for Testing and Materials (ASTM). The Air Transport Action Group (ATAG) reports that new alternative fuels produced from sustainable feedstock (e.g., algae, jatropha, domestic waste, or captured factory waste gases) deliver 80% of lifecycle carbon savings relative to traditional jet fuels.[z] Feedstock selection is important as it minimizes the impact to land and water use. Since its approval in 2011, the International Air Transport Association (IATA) reports that airlines have made more than 200,000 passenger flights using sustainable aviation fuels as shown in Fig. 24.

---

[z] Air Transport Action Group (ATAG) Facts and Figures. https://www.atag.org/facts-figures.html.

Year	Commercial flights operated on sustainable aviation fuels
2008	1
2013	3,000
2018	100,000
2020 (target)	1,000,000

Fig. 24. Commercial flights operated on sustainable aviation fuels.

Source: "10 years of flying with Sustainable Aviation Fuels", International Air Transport Association. https://www.iata.org/contentassets/92ddad17a72e45ae8fe2430e 38f65c79/saf10-infographic.pdf.

There are currently 5 global airports that support bio-fuel operations and 11 sustainable aviation fuel production facilities (a mix of in-production, under construction or in the final stages of securing financing). Biofuel requires no changes to airplanes or engines and performs as well as (or better) than Jet A/A-1. IATA is aiming for one billion passengers to fly on flights powered by a mix of jet fuel and sustainable aviation fuel (SAF) by 2025.[aa]

The FAA has complemented the CAAFI's efforts with the establishment of the FAA's Center of Excellence for Alternative Jet Fuels and Environment (ASCENT), a coalition of 16 leading US research universities and over 60 private sector stakeholders committed to reducing the environmental impact of aviation. A key focus is to enable production of sustainable aviation fuels at a commercial scale.[ab]

6. Bridging the Gap between Research and Technology Development and Product Insertion

In the early 1970s, NASA developed a methodology to assess the maturation of new technologies to inform investment decisions for continued development and adoption to new spacecraft design. In 1995, NASA expanded the Technology Readiness Level (TRL) scale as shown below. The DoD and the European Union adopted the TRL scale in the mid-2000s.[15]

Maturing technology is required but not sufficient to be successfully inserted into a commercial product. Through an iterative design, test and evaluation process, in partnership with industry, academia and research

---

[aa] https://www.iata.org/en/pressroom/pr/2018-02-26-01/.
[ab] https://ascent.aero/.

organizations, the technology maturation process must address early in the process the following critical questions and drivers:

- What is possible? (TRL 1–3)
- Can it be done? (TRL 3–5)
- Can it be done reasonably? (TRL 6)
- Will it work for this product application? (TRL 7–8)
- Does it pass the produce-ability, operability, and stakeholder accept-ability criteria?

One of the most treacherous stages in the maturation of new technology (to enable product adoption into new vehicle designs) is achieving TRL 6, which is the stage that attempts to transition a prototype component tested in a relevant environment to a commercial product. This transition is often called the "valley of death". Within the target relevant market and industry requirements and expectations, technology development must address the following "ilities" early on to enable product insertion.

- **Manufacturability** addresses the produce-ability of the vehicle, at high quality and rate.
- **Integratability** addresses interfaces and interactions between technologies and vehicle systems throughout the product's manufacturing system and overall operational environment.
- **Affordability** addresses the net value proposition, i.e., the favorable cost and benefit within the availability of resources.
- **Certifiability** addresses the acceptable path to certification approval with the relevant regulatory agencies, both within the US and the international community.
- **Reliability** addresses in-flight operations as well as airplane availability and operational dispatch.
- **Sustainability** addresses the lifecycle's long-term economic, social and environmental impacts on present and future needs as well as the uncertainties associated with people, climate change impacts and profit margin.
- **Maintainability** addresses the cost (labor, ease of access and time) of conducting inspections, servicing, repair and overhaul.
- **Operability** addresses in-use aircraft operations and compatibility with the airport infrastructure, pilot training and air traffic management systems.

- **Stakeholder acceptability** addresses the willingness of financial investors, aviation original equipment manufacturers, regulatory agencies, airspace operators and the public to take business and technical risks, including disruptive airport infrastructure development and aircraft production processes and operational procedures.

7. Global Aviation Environmental Research and Development[ac]

The 1990s saw a rapid increase in dialog around carbon dioxide emissions and climate change, resulting in the adoption of the United Nations Framework Convention on Climate Change (UNFCCC) in 1992, following the formation of the Intergovernmental Panel on Climate Change (IPCC) in 1988. Commercial aviation gradually came under more scrutiny, despite it contributing only a small 2% of global carbon dioxide emissions. By the late 1990s, some governments began to formulate plans to confront the problem. Figure 25 highlights the breadth of international efforts to address energy efficiency, emissions and climate change.[ad]

7.1. *European Union*

The European Union (EU) was at the forefront of international efforts, leading the formation of a committee comprised of industry leaders, academics and politicians. Their recommendations, which are published in *European Aeronautics: A Vision for 2020* with the overarching goal of "meeting society's needs and winning global leadership", resulted in the formation of the Advisory Council for Aeronautics Research in Europe (ACARE). ACARE was launched at the Paris Air Show in June 2001 and attracted over 40 member organizations and associations including representatives from the Member States, the European Commission and stakeholders from the manufacturing industry, airlines, airports, service providers, regulators, research establishments and academia.

The goal for ACARE was to develop and maintain a Strategic Research Agenda (SRA) that would help achieve the goals of Vision 2020.

[ac] https://www.icao.int/environmental-protection/Pages/EnvReport10.aspx.

[ad] International Coordinating Council of Aerospace Industries Associations (ICCAIA) chart referenced in 2010 ICAO Environmental Report https://www.icao.int/environmental-protection/Documents/EnvironmentReport-2010/ICAO_EnvReport10-Ch2_en.pdf.

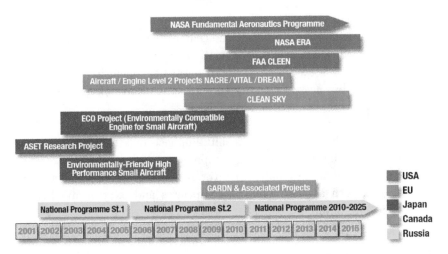

Fig. 25. National and regional research programs worldwide (2001–2015).

Source: ICAO environmental report 2010: aviation and climate change, International Civil Aviation Organization, 2010, Chapter 2. https://www.icao.int/environmental-protection/Documents/EnvironmentReport-2010/ICAO_EnvReport10-Ch2_en.pdf.

This approach has resulted in far-reaching influence on the funding and integration of aeronautical research in the EU.

The strategic agenda is integrally tied to the EU's collaborative research in Aeronautics and Air Transport, including the European Commission's Framework Programs (FP6, FP7 and Horizon 2020), Clean Sky Joint Technology Initiative, Single European Sky (SESAR) ATM Joint Undertaking, and national programs in many member states, research establishments as well as private company programs.

This research agenda began under Frameworks 6 and 7 but developed significantly with the separate "Clean Sky" program in 2008 with €1.6B over 5 years. Clean Sky was extended to "Clean Sky 2" in 2014 and folded back into the framework program in Horizon 2020, with €4B over 7 years. The centerpieces of Clean Sky are the Open Rotor engine program and the Breakthrough Laminar Aircraft Demonstrator in Europe (BLADE) laminar flow research programs. However, there are hundreds of smaller programs that cover the entirety of technology domains relevant to aviation.

When it became apparent that 2020 was too close a date to accomplish all of the desired goals, the committee formulated a new vision document "Flightpath 2050", which was released in March 2011, and remains the guiding document today. One feature of the EU's aeronautics agenda is that it puts industrial leaders at the center of the research programs — they lead the development and execution of the research agenda. The principles of EU research programs also reward the inclusion of small and medium sized businesses, thereby providing a means of growing these companies and broadening the industrial base.

The flexibility of the EU Framework programs allows each call for proposals (roughly twice a year) to shift emphasis over time because some projects stall while others make great progress, and other fields emerge. This has resulted in an ability to pivot recently to promising emerging technologies, such as thermoplastic composites and electric and hybrid-electric propulsion architectures.

7.2. *China/Russia*

Chinese investment in aviation research is difficult to estimate. The aerospace portion of China's Heavy Equipment Manufacturing Sector is estimated to be ~$3 billion of a total $7 billion, which includes rail, ships, etc. The Commercial Aircraft Corporation of China (COMAC) and the Aviation Industry Corporation of China (AVIC) have hired significant numbers of western aviation executives to assist with airplane development technology.

The China Russia Commercial Aircraft International Corporation opened its Shanghai office for the CR629 project in 2017. They are now jointly developing a wide body passenger aircraft, the CR929. The project includes the development of a new engine and is expected to take about 10 years to complete. From 2025 to 2028, first flights are expected and some systems, such as the landing gear have already been designed. Its target markets include China, Russia and other Asia-Pacific countries. Russia also continues to collaborate with other countries to advance its technological expertise.

With the rising demand for commercial aircraft, there are some new production programs emerging from outside the US and Europe, especially from China and Russia.

Though these new entrants could become challenging in the longer term, there are many hurdles they would need to cross — procurement of orders (from global airlines) and certifications from world-wide regulators, but most importantly, establishing and demonstrating a safe and reliable track record before they are widely accepted.[16]

Track Record of Significant Progress

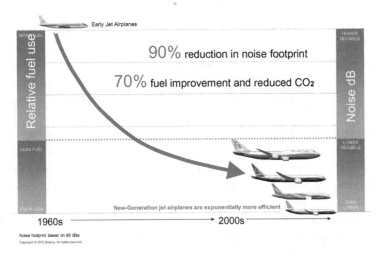

Fig. 26. Fuel burn use and noise footprint from 1960s to the 2000s.

Source: Boeing Commercial Airplanes and the Environment http://www.boeing.com/
resources/boeingdotcom/principles/environment/pdf/Backgrounder_Boeing_
Environment.pdf.

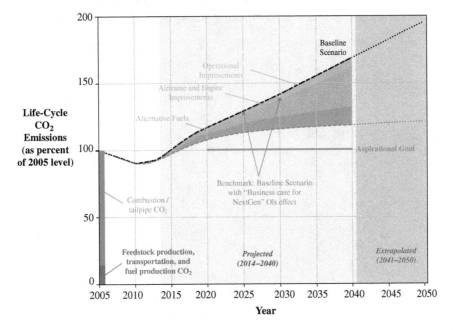

Fig. 27. Projected lifecycle carbon dioxide emissions impacts on the aggressive system
improvement scenario.

8. Conclusions

To date, commercial aircraft market pressures and the competitive land-scape have driven significant improvements in aircraft performance and noise reductions (as shown in Fig. 26), with average aircraft fuel burn and carbon dioxide over the last 60 years reduced by 70% and the noise foot-print around any given airport reduced by 90% (see footnote aa). However, the rapid growth in global air travel makes it a challenge to keep aviation carbon dioxide emissions at or below its 2005 levels.

There is reason to believe that carbon neutral growth at 2020 levels is possible. Figure 27 shows one strategy and even greater reductions can be achieved by using more ambitious technologies (e.g., lighter than air, autonomous traffic/UAS, eVTOL/hybrid-electric vehicles, etc.) as discussed in Appendix B.

Meeting these challenging goals will require a sustained level of com-mitment from both the US and other members of the international com-munity, the public and aviation industry sector as well as a global policy framework that will enable the implementation of the CAEP's 2% annual aviation energy reduction goal. Together with the CAEP/11's carbon dioxide emissions standards and an executable, socioeconomic CORSIA market offset plan, the aviation transport sector can achieve its environmental goals while making air travel safer, quieter and less expensive.

Appendix A. NASA Technology Readiness Level Definitions

Technology Readiness Level (TRL) Definitions

TRL	Definition	Hardware Description	Software Description	Exit Criteria
1	Basic principles observed and reported.	Scientific knowledge generated underpinning hardware technology concepts/applications.	Scientific knowledge generated underpinning basic properties of software architecture and mathematical formulation.	Peer reviewed publication of research underlying the proposed concept/application.
2	Technology concept and/or application formulated.	Invention begins, practical application is identified but is speculative, no experimental proof or detailed analysis is available to support the conjecture.	Practical application is identified but is speculative, no experimental proof or detailed analysis is available to support the conjecture. Basic properties of algorithms, representations and concepts defined. Basic principles coded. Experiments performed with synthetic data.	Documented description of the application/concept that addresses feasibility and benefit.
3	Analytical and experimental critical function and/or characteristic proof of concept.	Analytical studies place the technology in an appropriate context and laboratory demonstrations, modeling and simulation validate analytical prediction.	Development of limited functionality to validate critical properties and predictions using non-integrated software components.	Documented analytical/ experimental results validating predictions of key parameters.

(*Continued*)

(Continued)

Technology Readiness Level (TRL) Definitions

TRL	Definition	Hardware Description	Software Description	Exit Criteria
4	Component and/or breadboard validation in laboratory environment.	A low fidelity system/component breadboard is built and operated to demonstrate basic functionality and critical test environments, and associated performance predictions are defined relative to the final operating environment.	Key, functionally critical, software components are integrated, and functionally validated, to establish interoperability and begin architecture development. Relevant environments defined and performance in this environment predicted.	Documented test performance demonstrating agreement with analytical predictions. Documented definition of relevant environment.
5	Component and/or breadboard validation in relevant environment.	A medium fidelity system/component brassboard is built and operated to demonstrate overall performance in a simulated operational environment with realistic support elements that demonstrates overall performance in critical areas. Performance predictions are made for subsequent development phases.	End-to-end software elements implemented and interfaced with existing systems/simulations conforming to target environment. End-to end software system, tested in relevant environment, meeting predicted performance. Operational environment performance predicted. Prototype implementations developed.	Documented test performance demonstrating agreement with analytical predictions. Documented definition of scaling requirements.
6	System/sub-system model or prototype demonstration in an operational environment.	A high fidelity system/component prototype that adequately addresses all critical scaling issues is built and operated in a relevant environment to demonstrate operations under critical environmental conditions.	Prototype implementations of the software demonstrated on full-scale realistic problems. Partially integrate with existing hardware/software systems. Limited documentation available. Engineering feasibility fully demonstrated.	Documented test performance demonstrating agreement with analytical predictions.

7	System prototype demonstration in an operational environment.	A high fidelity engineering unit that adequately addresses all critical scaling issues is built and operated in a relevant environment to demonstrate performance in the actual operational environment and platform (ground, airborne or space).	Prototype software exists having all key functionality available for demonstration and test. Well integrated with operational hardware/software systems demonstrating operational feasibility. Most software bugs removed. Limited documentation available.	Documented test performance demonstrating agreement with analytical predictions.
8	Actual system completed and "flight qualified" through test and demonstration.	The final product in its final configuration is successfully demonstrated through test and analysis for its intended operational environment and platform (ground, airborne or space).	All software has been thoroughly debugged and fully integrated with all operational hardware and software systems. All user documentation, training documentation, and maintenance documentation completed. All functionality successfully demonstrated in simulated operational scenarios. Verification and Validation (V&V) completed.	Documented test performance verifying analytical predictions.
9	Actual system flight proven through successful mission operations.	The final product is successfully operated in an actual mission.	All software has been thoroughly debugged and fully integrated with all operational hardware/software systems. All documentation has been completed. Sustaining software engineering support is in place. System has been successfully operated in the operational environment.	Documented mission operational results.

Appendix B. Advanced Airframe and Propulsion Technologies

An overview of the airframe and propulsion technology improvements (including alternative fuels, which is on the horizon) that can revolutionize the landscape of future energy efficient commercial subsonic aircraft will be described later, as well as the technology insertion challenges that are being addressed through public-private partnerships to bring these aviation technology innovations to products.

To gain a first principle understanding of what drives the fuel consumption of an airplane, it helps to examine the Breguet Range Equation shown as follows:

$$\text{Aircraft Range} = \frac{\text{Velocity}}{\text{TSFC}}\left(\frac{\text{Lift}}{\text{Drag}}\right)\ln\left(1+\frac{W_{fuel}}{W_{PL}+W_{OEW}}\right), \quad \text{(B.1)}$$

where Drag is the aircraft drag (lbf), Lift is the lift produced by the aircraft (lbf), TSFC is the thrust specific fuel consumption ([lbf/hr]/lbf), Velocity is the aircraft velocity (nm/h), W_{fuel} is the fuel weight (lbf), W_{PL} is the payload weight (lbf), W_{OEW} is the operating empty weight (lbf).

The Breguet Range Equation shows that if the speed and TSFC (an engine parameter) are constant, the lift to drag ratio (L/D) must be maximized and the operating empty weight (W_{OEW}) must be minimized in order to fly a given payload as far as possible on a given amount of fuel. In cruise flight, Lift = Weight, therefore maximizing the L/D that is required for minimizing drag. Since the largest contributor to operating empty weight is the aircraft's structural weight, most efforts to reduce empty weight focus on reducing the weight of the aircraft's main structural components. The overall performance of the engine is to reduce the fuel burn per unit of delivered thrust. Aerodynamics improvements reduce drag and its associated thrust.

Table B.1 in Appendix B highlights the most promising propulsion and airframe technologies to enable the ICAO's fuel efficiency goal of 2% per year. Near team propulsion improvements for conventional tube and wing aircraft configurations are focused on increasing propulsor efficiency of the turbofan by reducing the Fan Pressure Ratio (FPR) to achieve improved fuel burn and reduced emissions and community noise levels. A lower FPR increases the fan diameter size to produce the same thrust, which results in the need for a shortened inlet (and lower-drag) and quieter nacelle design to prevent a weight and noise penalty associated with a larger fan

Table B.1: Promising technologies to enable ICAO fuel efficiency goals.

Propulsion architecture	Advanced Turbofan
	Geared Turbofan
	Open Rotor
Improved propulsive efficiency	Improved global design through advanced analytical tools
	Optimized components through advanced analytical tools
	Reduced nacelle weights that displace the optimum toward a decreased Fan Pressure Ratio (FPR) and increased Bypass Pressure Ratio (BPR)
	Advanced composites:
	• PMC, phenolic matrix composite
	• MMC, metallic matrix composite
	• CMC, ceramic matrix composite
	• Composite frames
	• Advanced alloys
	• Increased loading
	• Manufacturing technology
	• Combined feature functionality
	• Integrated installations
	• Zero hub fan
Improved thermodynamic efficiency	Improved materials to allow higher temperatures
	Improved compressor and turbine with 3D aerodynamics, blowing and aspiration
	Active tip clearance
Aerodynamics	
Viscous drag (friction drag) reduction	Riblets
	Active turbulence control
	Natural laminar flow
	Hybrid laminar flow control

(Continued)

Table B.1: (*Continued*)

Non-viscous drag reduction	Increased wing span (increased aspect ratio)
	Improved aero tools
	Excrescence reduction
	Variable camber with new control surfaces
	Morphing wing
Airframe and Engine Weight Reduction	Optimization of geometry through
	• Reduction of loads (active smart wing)
	• Bew joining process (removal of riveting) by
	• Stir-welding process
	• Super-plastic forming
	• Diffusion welding
	• Laser beam welding
	Metallic technologies
	• Al-Li
	• Al-Mg-SC
	• Advanced alloy
	Composite technologies
	• PMC
	• Fluoro-polymers
	• ARALL (aramid-fiber-reinforced)
	• CentrAl (alternating aluminum and fiber metal laminate)
	Multifunctional materials/Structures
	Nanotechnologies
	Health monitoring
Other Concepts	Electric landing-gear drive
	More Electric Aircraft (MEA)
	Fuel Cells

diameter. Improvements in the thermal efficiency of a turbofan engine are driven by the overall pressure ratio (OPR) and turbine entry temperature. The individual component-level efficiency from the propulsor, compressor and turbines, contribute to the overall engine efficiency. Although higher OPR (>60) increases the thermal efficiency of the engine and improves the engine's specific fuel consumption, there are significant challenges with a reduction in overall core size, and increased temperatures in the compressor can lead to compressor blade material failure. High temperature material development for engine components (e.g., the combustor, compressor, and turbine) as well as for airfoil cooling technology are required to enable increases in the OPR. Thermal efficiency improvements will require a combined approach while maintaining compressor and turbine efficiencies.

References

1. "The Growth in Greenhouse Gas Emissions from Commercial Aviation" Oct 2019, Environmental and Energy Study Institute. https://www.eesi. org/papers/view/fact-sheet-the-growth-in-greenhouse-gas-emissions-from-commercial-aviation.

2. https://www.icao.int/environmental-protection/Lists/ActionPlan/ Attachments/30/UnitedStates_Action_Plan-2015.pdf.

3. J. Scheelhaase, S. Maertens, W. Grimme and M. Jung, "EU ETS versus CORSIA — A critical assessment of two approaches to limit air transport's CO2 emissions by market-based measures", *Journal of Air Transport Management* **67** (2018) 55–62. doi:10.1016/j.jairtraman.2017.11.007. https://www.researchgate.net/publication/323488716_EU_ETS_ versus_CORSIA_-_A_critical_assessment_of_two_approaches_to_limit_air_ transport%27s_CO_2_emissions_by_market-based_measures.

4. M. Sinnett, "787 no bleed systems: saving fuel and enhancing operational efficiencies", *AERO*, 2007. https://www.boeing.com/commercial/aero magazine/articles/qtr_4_07/article_02_1.html.

5. H. Morris, "Airline weight reduction to save fuel: the crazy ways airlines save weight on planes", *Traveller*, 4 September 2018. http://www.traveller. com.au/airline-weight-reduction-to-save-fuel-the-crazy-ways-airlines-save-weight-on-planes-h14vlh.

6. M. Bijvank, M. Dobber, M. Soomer, *et al.*, "Planning drinking water for airplanes", 29 April 2005. https://www.win.tue.nl/~rhofstad/KLM_paper-NewVersion.pdf.

7. T. Dubois, Airbus "Product Responsibility: A lifecycle approach to improved environmental performance". https://www.airbus.com/company/sustainability/environment/product-responsibility.html.

8. T. Spence, "Old airplaces find an afterlife as recycled resource", *Euractiv*, 14 November 2012. https://www.euractiv.com/section/trade-society/news/old-airplanes-find-an-afterlife-as-recycled-resource/.

9. D.V. Zante and F. Collier, "The environmentally responsible aviation project", National Aeronautics and Space Administration, 2016. https://ntrs.nasa.gov/archive/nasa/casi.ntrs.nasa.gov/20170004101.pdf.

10. G.M. Bezos-O'Connor, M.F. Mangelsdorf, H.A. Maliska, *et al.*, "Fuel Efficiencies Through Airframe Improvements", 3rd AIAA Atmospheric Space Environments Conference, Honolulu, Hawaii, 27–30 June 2011. https://doi.org/10.2514/6.2011-3530.

11. O. Gur, J.A. Schetz and W.H. Mason, "Aerodynamic considerations in the design of Truss-braced-wing aircraft", *AIAA Journal of Aircraft* **48** (2011) 919–939. https://www.researchgate.net/publication/283010588_Aerodynamic_Considerations_in_the_Design_of_Truss-Braced-Wing_Aircraft.

12. T.P. Duffy and R.H. Jansen, "Turboelectric and hybrid electric aircraft drive key performance parameters", *2018 AIAA/IEEE Electric Aircraft Technologies Symposium*, Cincinnati, Ohio, 9–11 July 2018. https://ieeexplore.ieee.org/document/8552810.

13. C. Alcock, "Is urban air mobility part of business aviation's future?" *AINonline*, 18 October 2019. https://www.ainonline.com/aviation-news/business-aviation/2019-10-18/urban-air-mobility-part-business-aviations-future.

14. EPA, https://www.epa.gov/laws-regulations/summary-clean-air-act.

15. H. Mihaly, "From NASA to EU: the evolution of the TRL scale in public sector innovation", *The Innovation Journal* **22** (2017) 1–23. https://www.semanticscholar.org/paper/From-NASA-to-EU%3A-the-evolution-of-the-TRL-scale-in-H%C3%A9der/7d72b6a1dd7f1920934464d0075ea6aac8a848a3.

16. R. Lineberger, "2019 global aerospace and defense industry outlook", Deloitte, 2019. https://www2.deloitte.com/content/dam/Deloitte/global/Documents/Manufacturing/gx-eri-2019-global-a-and-d-sector-outlook.pdf.

https://doi.org/10.1142/9789811217869_0007

Chapter 7

Energy Demand Futures by Global Models[*]

O.Y. Edelenbosch[†,§] and Detlef P. van Vuuren[‡]

[†]*Department of Management and Economics, Politecnico di Milano,*
Via Lambruschini 4/B, Milan, Italy

[‡]*PBL Netherlands Environmental Assessment Agency, Bezuidenhoutseweg 30,*
2594 AV Den Haag, The Netherlands

[§]*oreane.edelenbosch@pbl.nl*

1. Introduction

Changes in energy demand play a major role in determining future carbon dioxide emissions at the global level. In order to analyze the complex interactions of different sectors contributing to global greenhouse gas emissions, scientific models have been developed that are referred to as integrated assessment models (IAMs). In the broadest definition IAMs describe the key processes in the interaction of human development and the natural environment to gain a better understanding of global environmental problems. Especially in the last two decades they have become common tools of climate change analysis and the number of scenarios has dramatically increased.[1] Many of these models address policy issues such as the mitigation measures needed to reach certain climate targets and cost-effective emissions reduction pathways. Their results have been used extensively by different types of global assessment, such as those done by the Intergovernmental Panel for Climate Change (IPCC).

---

[*]This chapter is a summary of O.Y. Edelenbosch's PhD thesis *Energy Demand Futures by Global Models: Projections of a Complex System* (Utrecht University, 2018).

1.1. *IAM projections*

IAMs are not a homogenous group of analysis tools. Instead, they are evolving from different model classes — they can differ in terms of their system boundaries, the amount of detail in representing the various system parts and their solution method.[1] Some models are characterized by relatively detailed biophysical processes and a wide range of environmental indicators; in contrast, other models have more details in their representation of economics and policy instruments, or more sectoral and technology detail.[2] Each approach has its own strengths and weaknesses. Key input categories of assumptions are demographic and economic development, lifestyle change, natural source availability, technology development and policy and government.

IAMs cannot — and are not intended to — be precise projections of future events. Such predictions are only possible when all variables and relationships in a system are known and the system can be observed in controlled and reproducible situations. Environmental problems are, however, characterized by complex relationships and a high level of uncertainty.[3] The longer the time horizon is, the larger the uncertainty. It has therefore often been stressed that IAMs are often used for insights but not for exact numbers.[4] A scenario analysis does not aim to show the most likely development, but assesses different pathways under key assumptions to distill robust insights and evaluate uncertainties.[5] They provide plausible descriptions of socio-economic, technological and environmental futures.[6] Using IAMs to assess climate change should therefore be seen as particularly useful for exploring possible trends in relation to uncertain development of driving forces. In this chapter we take a close look at the IAM demand sector projections. A comparison with "bottom-up technical potential" studies shows that there is more room for energy efficiency improvement than currently deployed in most IAM climate policy scenarios. The IAMs suggest that capturing these additional opportunities will require ambitious policy interventions.

1.2. *Scenario analysis*

A clear understanding of the critical assumptions made by each model and their uncertainties is essential for using these models correctly. Another approach to better understand the uncertainty associated with future projections is to compare the results of different models. Multi-model comparison builds on the diversity of models (in terms of their structural and parametric assumptions) to reveal areas of agreement and areas where there appears to be significant uncertainty. This is essential for understanding the potential benefits and liabilities of different approaches to mitigation.

The importance of multi-model comparisons and scenario development has grown rapidly in recent years and has led to works such as the *Shared Socio-economic Pathways* (SSPs) (see Box 1).[7,8] This scenario framework has been developed to cover the range of plausible future developments affecting climate change in a comprehensible (i.e., limited number of scenarios) and comprehensive manner (i.e., covering the space of plausible futures sufficiently). More specifically they intend to explore the consequences of socio-economic developments on anthropogenic climate change and the available response options through mitigation and adaptation.[9]

2. Energy Demand in IAMs

IAMs tend to focus analyses more on mitigation of energy supply side emissions, while relatively less attention has gone to the use of energy and the role of energy reduction in a global setting to achieve climate targets. However, more details, especially in the transport sector, have been recently added.[12] Generally, the energy supply sector is also represented with more detail in IAMs and energy system models than the energy demand sectors.

There are several key reasons for this:

- Energy demand sectors are highly diverse with many sub-sectors, different functions for which energy is used, technologies and users making these sectors more difficult to describe by models. Moreover, these different users vary in their preferences and needs.
- There is a faster turnover of capital stock and innovation cycles in the energy demand sector than in the energy supply sector, which adds to the sectors' complexity.
- The rules affecting future energy demand change are less defined, as actors use many more criteria in making decisions than the "rational" cost-optimization to investment decisions that are typically used in the energy supply sector.[1]

Box 1. Shared Socio-economic Pathways (SSPs)

IAMs are used to develop different types of scenarios. These scenario types correspond to different research questions.

- Baseline emission scenarios (often called "counterfactuals") of IAMs show futures where no explicit measures to reduce GHG emissions are taken. The baseline scenarios can still correspond to very different sets of assumptions. Often, storylines are introduced to ensure a consistent set of assumptions (e.g., for population and income and technology development).
- Mitigation scenarios are developed to look into the impact of policies. Such scenarios can start from existing policy formulation, but also explore the question on how to reach certain policy outcomes (e.g., normative targets for reducing emissions).[10]

Other characterization of scenario types can also be made, such as descriptive versus normative scenarios. Where it could justifiably be argued that all scenarios are normative (as they contain interpretations of developments), descriptive scenarios explore different future pathways while normative scenarios describe probable futures, which in some cases are referred to as the reference scenario.[11]

An example of descriptive scenarios is the recently published *Shared Socio-economic Pathways* (SSPs). The scenario framework consists of a set of five qualitative baseline scenarios of future changes in demographics, human development, economy, lifestyle, policies and institutions, technology and environment, and natural resources.[7] The scenarios differ in their emission mitigation and adaptation challenge due to varying reference emissions in the absence of climate policy as well as the projected "mitigative capacity" of the projected future society. SSP1 resembles a green growth world in which policies and behaviors are oriented towards a more sustainable development. This is translated into assumptions on rapid technological change directed toward environmental friendly purposes, lessened inequalities, educational and health investments, relatively low population growth and high land productivity. SSP3, the rocky rivalry scenario, forms in a way that is opposite of SSP1, where climate mitigation would require a much larger effort. In this scenario there are increasing concerns about national competitiveness and security, where also due to increased regional conflicts, countries focus mainly on domestic or regional issues. Moderate economic growth, rapid growing populations, slow technological change (especially in the energy sector), low investment in human capital, and inequality, characterize this scenario. SSP2 is the "middle of the road" scenario in which future social, economic and technological trends do not show distinct shifts from historic patterns. As such SSP2 forms the intermediate case between SSP1 and SSP3.[7]

To avoid all these complexities, modelers typically choose to represent energy demand in a very stylized way, i.e., describing energy demand as a function of activity levels, an elasticity of demand, and price elasticity. An indirect consequence, however, is that much less attention has been paid to the use of energy and the role of energy reduction in a global setting to achieve climate targets. However, as interest in model outcomes focuses more on concrete policies and measures, these aggregated descriptions become less useful and easy to interpret. More detailed information is, for instance, needed to support policies that look into questions like how the Paris Agreement can be implemented. Additional details also allow one to better relate models to sector specific studies examining current mitigation potential.

On the other hand, adding more detail also comes at a cost. For long-term global projections more detail does not imply greater accuracy. In fact, details can lose meaning over time as uncertainties increase. Given the heterogeneity of many demand-side processes it is not straightforward whether adding a more detailed representation would improve capturing the sectors dynamic behavior.

3. Cross-Model Demand Sector Comparison

In order to better understand the role of energy efficiency within global mitigation pathways, this chapter examines the projected global futures of the industry, transport and buildings sectors within different sets of integrated assessment models. The chapter is based on four papers that compare IAM demand sector projections with historic data and sector specific studies across the three largest demand sectors.

The chapter focuses on the following three questions:

- How do IAMs represent energy demand and what do they project?
- How do energy demand sectors in IAMs respond to climate policy?
- How do IAMs perform in their energy demand representation?

The three questions are addressed based on the combined results of all four papers (see Table 1).

Paper 1 assesses industrial energy demand and greenhouse gas emissions of eight IAMs. A descriptive questionnaire that addresses the assumptions made in the model's structure, system boundaries, energy and material demand drivers, technology change and policy measures has

Table 1. Overview of four key papers that are summarized in this chapter.

	Paper 1	Paper 2	Paper 3	Paper 4
Sectors	Industry	Transport	Transport	Buildings, Industry and Transport
Method	Qualitative questionnaire and output database	Qualitative questionnaire, output database and Laspeyres Decomposition	Calculated price and income elasticities from scenarios	Cross SSP comparison of Shapley–Sun decomposition
Scenario	Baseline (comparable to SSP2)	Baseline (comparable to SSP2)	Baseline (comparable to SSP2)	SSP1, SSP2 and SSP3 (see Box 1)
Participating models	AIM-CGE, DNE21+, GCAM, Imaclim-R, IMAGE, MESSAGE, POLES and TIAM-UCL	AIM/CGE, DNE21+, GCAM, GEM-E3, Imaclim-R, IMAGE, POLES, MESSAGE, REMIND, TIAM-UCL and WITCH	IMAGE, MESSAGE, POLES, REMIND, TIAM-UCL and WITCH	AIM/CGE, GCAM, IMAGE and MESSAGE-GLOBIOM
Validation	Cement sector specific analysis, WEO projection	Comparison to historical indicators	Comparison to historical indicators	Comparison technical bottom-up study
Reference	See Ref. 13	See Ref. 14	See Ref. 15	See Ref. 16

been constructed and filled in by all participating models. The detailed reports are presented in the Table A.1. The paper explores how differences in these model characteristics and assumptions affect the differing model outputs. A separate section of the paper looks specifically at the cement sector representation comparing the models' projected global emissions and material and energy consumption. In terms of GHG emissions the cement sector is one of the largest industrial sectors.

Paper 2 also explores how model inputs and design affect their outputs focusing on the transportation sector. It provides an overview of the key transport model characteristics, in terms of main drivers, dynamics and assumptions, of 11 IAMs that were collected through a qualitative questionnaire. In addition, the contribution of the sectors' activity growth, modal structure, energy intensity and fuel mix to the projected emissions pathways are quantified across the participating models. In order to do so the Laspeyres index decomposition method is used.[17] Results are compared between models, as well as scenarios with and without climate policy and against historical transport trends. The results of the stocktaking of this work can be found in the Tables A.2 and A.3.

Paper 3 explores how different models estimate the impact of a variety of fuel price scenarios in the transport sector, by calculating the implicit fuel price elasticities based on the model output. It provides a consistent, transparent framework to show how each model connects fuel prices to transportation energy demand. Through a similar approach it provides a clear way to compare the way transport demand to income elasticities are calculated in the models. Historical price and income elasticities have been studied extensively, which makes it possible to compare the model results to the empirical (historical) data.

Paper 4 covers all sectors and compares industry, transport and buildings demand futures in detail. It uses the Shapley–Sun[17] index decomposition method to explore how future changes in demand sector emission pathways are affected by population change, final energy per capita, electrification and fuel switching. This detail enables a comparison of the demand sectors with a sector specific technical assessment, published in the recent UNEP GAP report.[18] Not only are different models compared, so too are the effect of the different SSP baselines on demand sector developments and the required avoided emissions to meet stringent mitigation targets.

4. Global Demand Scenarios

4.1. *How do IAMs represent energy demand and what do they project?*

Most models show that in the absence of new policies, energy demand will continue to increase, driven by increasing population and global incomes. Most also show that significant demand reductions can be achieved with policies that, among other things, encourage investment in energy efficiency. Some models directly relate economic and demographic drivers to energy demand, while others focus on energy service demand, such as the demand for materials, industrial products or kilometers traveled. Some models are *more "bottom-up" — using a detailed representation of the energy system — and others are more "top-down" with more focus on using macroeconomic methods to explore the interaction between the energy system and the economy.*

Energy service demand in some models are specified per sub-sector, such as the demand for cars or bus transport, or in the industry sector, for example, cement and steel demand. If and which sub-sector division is made differs per model. The sub-sector shares, i.e., the structure of the sector, are either set exogenously over time or respond to price or saturation constraints, or in some transport models travel time. Most models include a representation of current and future technologies to fulfill the required service demand. But as in this case the level of detail differs. The technologies then compete on the basis of relative costs, leading to energy efficiency improvements or fuel switching when fuel prices increase. In some cases, technology development is driven by exogenous assumptions while in other cases it is through learning by doing functions. An important difference across the models is the solution type used. The different models analyzed here include intertemporal optimization models as well as simulation models.

Figure 1 shows the wide range of industry energy projections resulting from the assumptions models make on the development of technology, demographic changes, policy, lifestyle changes, structural changes and natural resource availability. It is obvious that the longer the period of the projection is, the greater the differences between the models will be. Sharp differences in assumptions about behavior (e.g., will driving continue to grow?), the rate of technology advancement and policy stringency, are magnified by the passage of time. An example can be seen in Fig. 2, which illustrates the range of assumptions made about electric vehicles cost development in different models.

In SSP2 (the middle of the road scenario, see Box 1), direct carbon emissions are projected to increase in 2050 by –0.4–2.9, 1.2–6.7 and 3.3–8.2 Gt carbon dioxide for buildings, industry and transport sector, respectively. In SSP1, which emphasizes a more sustainable development, lower carbon emissions are projected. In SSP3 (rocky rivalry scenario, see Box 1) industrial emission are higher. By use of the decomposition method the different developments leading to the higher emissions have been compared across models and different scenarios for all three demand sectors. Here we

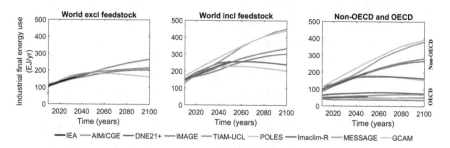

Fig. 1.   Baseline final energy demand projections in the industry sector up to 2100: (a) Global excluding feedstock, (b) global including feedstock, and (c) non-OECD and OECD countries including feedstock. (Reference: Paper 1, see Table 1.)

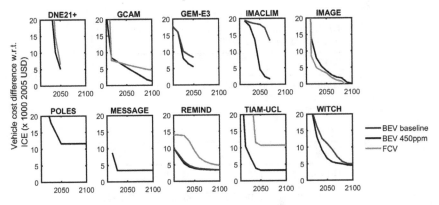

Fig. 2. Difference in USA LDV fuel cell vehicle (FCV) and battery electric vehicle (BEV) investment costs compared to conventional vehicles (ICE) in the mitigation scenario. POLES and MESSAGE FCV investment cost remain >20.000 2005 USD more expensive than conventional vehicles, i.e., outside the displayed range. (Reference: Paper 2, see Table 1.)

distinguish between population growth, final energy use, electrification and the direct emissions of non-electric fuels. The decomposition analysis shows that in baseline projections key different assumptions about population growth and growth of final energy per capita over the coming decades are largely responsible for differences in estimates of future transport, industry and buildings carbon emissions. Table 2 show the final energy demand ranges in the buildings, industry and transport sector in 2010, and its estimated range by 2050 and then 2100. There is already disagreement about the baseline energy consumption assumed for 2010. Figure 1 shows significant differences for industrial energy use, particularly feedstock. This is a non-trivial issue since feedstocks represent 17% of industrial energy use. The difference across the models emphasizes the importance of clear boundary definitions when comparing model projections. The alternative baseline scenarios SSP1 and SSP3 strongly influence the final energy consumptions and the required emission reduction to meet a specific climate target. In an SSP1 world, policies toward a more sustainable development are assumed, including rapid technology development and more environmental preferences such as greater use of public transport with higher load factors and more reuse and recycling. SSP3 on the other hand is characterized by a strong population growth, slow technology development and regional conflicts, leading countries to focus on regional issues and less on collaboration.

Table 2.   Range of global final energy demand in
EJ in the buildings, industry and transport sector
in 2010, 2050 and 2100.[19]

2010			
Buildings	120–130		
Industry	120–140		
Transport	90–100		

2050	SSP1	SSP2	SSP3
Buildings	170–190	180–220	190–240
Industry	150–270	190–240	210–300
Transport	120–190	160–190	150–230

2100	SSP1	SSP2	SSP3
Buildings	160–250	240–360	200–370
Industry	120–280	210–350	300–460
Transport	110–180	200–300	200–360

Note: The values are rounded off to the nearest 10.
(Reference: Paper 4, see Table 1.)

**Growth of global industrial energy demand is mostly determined
by developments in non-OECD countries, while demand in
OECD countries remains more or less constant.** The industrial
energy demand in 2100 in non-OECD regions ranges across model projections from 150 to 400 EJ.[a] Paper 1 (see Table 1) shows that regional
developments are at the root of global developments. The industrial
energy intensity[b] annual reduction rates of non-OECD countries range
from 1.8% to 2.2%. This is historically significantly faster and would
require these regions to converge more to patterns seen in OECD countries
in the past. While the models agree on this, the range still leads to
large differences in final energy consumption in the long run (see Fig. 1).
The models differ strongly in the detail of activity and technology representation of the industry sector. However, the results did not show any
systematic different between models with or without sector detail.

In the transport sector the differing levels of energy demand
for passenger travel can be traced back to the projected travel

---

[a]The values are rounded to the nearest 10.
[b]This is equal the energy use per GDP.

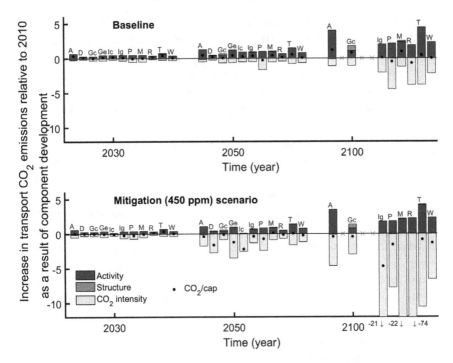

Fig. 3.   Passenger transport direct carbon dioxide emission increase relative to 2010 due to activity, structure or carbon dioxide intensity development, in accordance to a single factor change for baseline (top) and mitigation (bottom) scenarios in AIM/CGE (A), DNE21+ (D), GCAM (G), GEM-E3 (Ge), Imaclim-R (Ic), IMAGE (Ig), POLES (P), MESSAGE (M), REMIND (R), TIAM-UCL (T) and WITCH (W). WITCH and REMIND show results at the LDV level. (Reference: Paper 2, see Table 1.)

volume and the expected magnitude of energy efficiency improvements. Baseline energy transport demand varies: the energy use in 2050 was from 93 to 121 EJ and in 2100 from 130 to 206 EJ (for comparison: the energy use in 2010 was 47–55 EJ). Although the annual growth rates of passenger travel cover a relatively small range compared to what has been observed historically, this can lead to a significant spread in projected demand (see Fig. 3). Transport demand increase across the modes ranges from a factor two to a factor five. In physical terms LDV global demands ranges from traveling 68.000 to 123.000 billion passenger kilometers in 2100, compared to on average 22.000 billion passenger kilometers[c] in 2010, clearly affecting the energy requirements.

---

[c]Average value across models with a spread of 17.000–26.000 billion kilometers in 2010.

Projections of energy demand in industry can benefit from modeling specific subsectors of the industry sector, since industrial subsectors can behave differently in response to climate policy. Since it is a major energy consumer, several models examine cement production in some detail. Across these IAM cement models, Paper 1 (see Table 1) shows that the specific energy consumption (GJ/tonne product) for cement and clinker making is generally projected to decline driven by technology development. Literature suggests that the energy use for clinker making can drop to 2.9 GJ/tonne clinker and when improved equipment for cement making and lower clinker to cement ratios are used the energy use could drop further to 2.1–2.7 GJ/tonne cement.[20] The baseline model projections are well within the technical potential. In fact, considerable improvement of the energy efficiency would still be possible in the mitigation scenarios compared to the baseline projections. Modeling industrial energy service demand could in addition provide the opportunity to relate the consumption of materials to non-economic drivers such as infrastructure or buildings stock development and possibly better evaluate demand saturation scenarios.

4.2. *How do energy demand sectors in IAMs respond to climate policy?*

Energy efficiency and fuel switching (including electrification) both play an important role in emission mitigation. In the short term, both are important though fuel switching fuel switching is more dominant in the long run (see, for example, the emission reduction due to carbon dioxide intensity change in the transport sector in Fig. 3). There are several strategies to mitigate demand sector emissions that can be categorized at a higher level into: (1) increasing energy efficiency, (2) changing fuel mix, and (3) reducing or changing energy service demand (i.e., kilometers driven, floor space requirements, steel production). This categorization is used to discuss the projected mitigation pathways in the demand sectors.

4.2.1. *Energy efficiency*

The mitigation scenarios of SSP1, SSP2 and SSP3 show energy efficiency improvements in all demand sectors. The projected required energy efficiency improvements are within the energy efficiency potential that was estimated based on an in-depth review of numerous technical studies for each sector. In the middle of the road SSP2 scenario this leads to a 0.4 (0.0–0.7), 1.1 (0.2–2.0) and 1.3 (0.0–2.5)[d] Gt carbon dioxide reduction in the buildings, industry and transport sectors, respectively, in 2030. The IAM sector results are compared to a detailed technology-oriented assessment that was performed recently by UNEP Gap to calculate sectoral emission reduction potentials. These reduction potentials are with respect to an average emission intensity in 2030 that was estimated by the World Energy Outlook.[21] Although the IAM reductions are well within the potential estimated by a sector-by-sector literature review, these bottom-up estimates performed by for the same sectors, show that there is still significantly more room for energy efficiency improvements (see Table 3).

The average passenger transport energy efficiency decreases in a no climate policy scenario to 0.5–1 MJ/passenger kilometer in 2100. The inclusion of energy service demand projections enables us to compare projected energy efficiency improvements to those estimated by bottom-up studies (see Fig. 4). The model projections show that energy efficiency improvements are an important factor to decrease emissions from passenger transport. In 2100 the energy used

Table 3. Comparison of average 2030 avoided emissions in the IAMs Under a SSP2 2°C pathway in Gt CO_2 with the emission reduction potentials found in the sector-by-sector analysis bottom-up analysis.

	Buildings	Industry	Transport
Integrated assessment models	0.7 (0.3–1.0)	2.6 (0.9–3.2)	1.7 (0.9–2.7)
Efficiency	0.4 (0.0–0.7)	1.1 (0.2–2.0)	1.3 (0.0–2.5)
Electrification	0.0 (−0.1–0.0)	0.0 (0.0–0.2)	0.0 (0.0–0.3)
Fuel switch	0.3 (0.2–0.4)	1.4 (0.6–1.9)	0.3 (0.0–0.8)
Technology-oriented assessment	—	—	—
Efficiency	1.2–1.8	1.6–2.8	3.0–4.9
Electrification	—	—	—
Fuel switch	0.4–0.9	0.4–0.6 + 0.9–1.5 (CCS)	0.6–0.8

Note: The negative sign in the ranges indicates increased emissions instead of avoided. (Reference: Paper 4, see Table 1).

---

[d]On average with between brackets average model range.

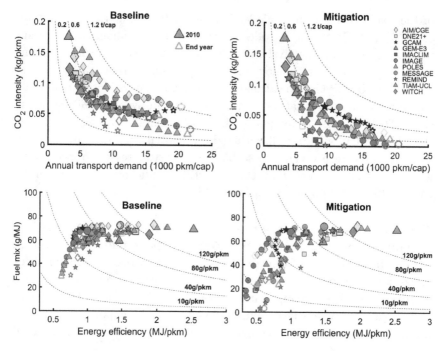

Fig. 4. (Top) Global passenger transport activity per capita (x-axis) compared to carbon dioxide intensity (y-axis) development over time. The start and stop years are indicated with the marker, and each marker indicates a 10-year time step. DNE21+ and GEM-E3 model projections run to 2050, Imaclim-R to 2070 and the rest until 2100. The carbon dioxide emissions per capita are indicated by the plotted isolines. The left panel shows the baseline and right mitigation scenario. (Bottom) Global passenger transport energy intensity (x-axis) compared to fuel mix (y-axis) development in top figures. The isolines indicate emissions per passenger kilometer. (Reference: Paper 2, see Table 1).

per passenger kilometer improves 46–72% without any climate policy. This is further increased to a significantly lower level through further technology improvement and the adoption of alternative propulsion techniques such as electric vehicles, as response policy. From a bottom-up technical perspective in 2030, fuel consumption could be reduced by 30–50%, while switching to alternative driving mechanisms like electric vehicles could reduce fuel consumption even further. Although switching to more energy efficient modes, such as taking the train, could also contribute to reduced energy use, the IAM modal shares remain close to current modal shares or switch limitedly to faster, less energy efficient modes such as cars and planes. The estimated energy efficiency improvements could even be larger if modal shift policies were included as well.[22]

4.2.2. *Fuel switching*

Fuel switching is central to meeting stringent climate targets and is increasingly important over time. This constrains the types of energy efficiency technologies appropriate for low-carbon scenarios (e.g., efficiency of petroleum-based systems is relevant only if a low-carbon substitute for petroleum is used). The global scenarios show that switching to low carbon fuels is the most effective strategy to mitigate emissions in the energy demand sectors. There are different pathways to do so, depending on the sector, energy service requirements and technology options. Paper 4 shows that in the buildings sector the electrification trends seen in the last decades are projected to reach 43–49% in 2050 and 57–69% (model range) in 2100 under no climate policy assumptions. These would increase further to 45–53% in 2050 and 73–93% in 2100 in response to climate policy. This means that the largest share of the buildings sector emissions are not emitted directly but indirectly during the electricity production. The cross sector perspective of integrated assessment models is particularly suitable to analyze the effectiveness of fuel switching to decarbonize emissions. In the near term, incentives to switch fuels can be limited because electricity, for example, may still be carbon intensive, while term energy reductions will have less effect in the longer run if fuel sources such as electricity and liquid fuels have been decarbonized. Indeed, we find that fuel switching (as a measure to mitigate emissions) in the demand sectors increases over time.

Models show different fuel switching strategies in the passenger transport sector. The passenger transport comparison shows different pathways of fuel transition, with different fuel types being deployed and different rates of deployment. The uncertainty of technology development is reflected in the range of electric as well as fuel cell vehicle capital cost projected by the models (see Fig. 2), but does not fully explain the different choices made by the models. Other important aspects are: fuel price, non-financial factors or calibration factors, and model solution methods shown by the price elasticity analysis. The transport sector has in recent decades been dominated by oil use but large-scale decarbonization of fuel is required to meet climate targets (see Fig. 4). In some projections, market shares of the passenger transport fuel need to increase to 80% of the electricity or hydrogen, or reach a global 50% of the biofuels by the end of the century, which would imply a clear break with historical trends.

Similarly, switching from coal to electricity is an important measure to reduce emissions in the industry sector. Interestingly, models that explicitly include industrial technologies seem to be less flexible in switching to alternative fuels. In the industry sector there is a reasonably high agreement of future fuel shares in the baseline, remaining close to current shares. Most models project a slight increase in electricity use and a decrease in fossil fuel use, both between 10% and 20% change, while the fuel share responds strongly in a 2°C scenario (see Fig. 5). Mitigating emissions occurs through a combination of fuel switching and energy intensity improvements. In all models, the percentage reduction in industrial final energy use remains fairly constant after 2040, as compared to baseline, although it spans a range of 10–50%. In this set of models, and a similar division is seen in the price elasticity paper, the more technology-detailed models are less flexible to fuel switching and see more potential in energy efficiency improvements. An issue here could be that the energy efficiency improvements are constrained to current knowledge of technology developments. Some industrial processes are more suited to apply carbon capture and storage measures to reduce emissions. The representation of this in the global models has however not been studied specifically during this research.

4.2.3. *Demand for energy services*

The models project that demand for energy services is not as responsive to climate policy. The potential of policy to affect demand is not well understood and therefore difficult to evaluate with the models. The earlier discussion showed that not all IAMs

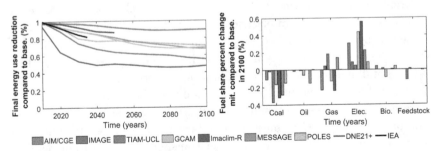

Fig. 5. (Left) Mitigation scenario final energy demand as a portion of the baseline scenario final energy demand. (Right) Percent change in fuel share mitigation scenario compared to baseline. (Reference: Paper 1, see Table 1.)

explicitly account for energy service demand in their projections. In the cases where energy service levels were reported, such as for most of the transport sector models, the models examined find that energy service demand is hardly responsive to climate mitigation policy, in part due to the fuel price elasticity of demand being comparatively low in this sector. The common way to represent climate policy is through a carbon tax affecting fuel price. In some models service demand is related only to exogenous GDP assumptions and therefore cannot be affected by fuel price change. In other models, alternative mitigation measures such as fuel switching are more attractive because they are less reliant on behavioral change. Place-based transport studies emphasize the mitigation potential of infrastructure and behavioral change especially in the urban environment, leading also to local co-benefits. These studies indicate that the cost optimization perspective of IAMs might underestimate the potential of energy service demand change, which could complement the radical fuel switching required in current projections.[22] The SSP1 scenario results (see Table 2) show that sustainable lifestyles would strongly improve the mitigation potential in the demand sectors.

4.3. *How do IAMs currently perform in their energy demand representation?*

One way to test the reliability of models used for future projections is to compare the estimated trends to historic indicators. While tests involving all three demand sectors were performed in Paper 4 (and can also be found in the broader literature), the following discussion will focus on transportation as an example.

Projected future trends in the transport energy demand sector are generally comparable to historical indicators in activity growth, modal shift, energy intensity, energy and income elasticities and prices elasticities. The conclusion of these tests is that the future projections of the transportation models do a good job of remaining within the historical energy demand and energy intensity ranges. In fact, the variation reported historically in this set of countries is larger than the range across models. Fuel switching projections however go beyond historical measurements. The transport sector has for the last decades been largely (>90%) dependent on oil while some projections show significant shifts to electricity and alternative fuels.

Activity growth, energy intensity, modal shift and fuel mix developments contributing to the projected greenhouse gas emissions are untangled through the Laspeyres index decomposition analysis. The same method has been used in energy research in recent decades to understand historical trends of these factors. Fuel mix was not examined historically as this remained more or less dominated by oil. While energy intensity in all models decrease, this has historically not always been the case. All models examined show energy intensity decreasing but mitigation scenarios show the fastest decrease.The mitigation scenarios are close to the most rapid energy intensity decrease documented in recent history.

Looking to the future, LDV energy demand elasticities to oil and gas prices are projected to range from −0.2 to −0.5 in 2030, close to the range described in the empirical literature. In the very long term (30–40 years), LDV energy demand elasticity values vary between −0.4 and −2.1, showing either continuous demand or increased demand responses over time (see Fig. 6). The energy demand to income elasticity values range between 0.3 and 1.4. This is within the range reported in the literature. Key model drivers

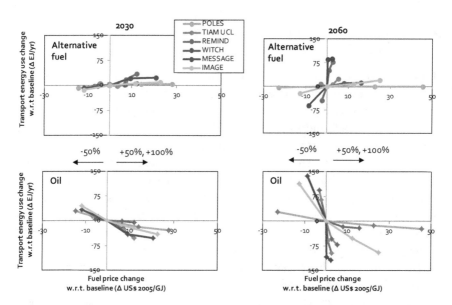

Fig. 6. The oil (bottom row) and alternative fuel (AF) (top row) energy demand response to −50%, +50% and +100% oil and gas price shocks in 2030 (left column) and 2060 (right column). Alternative fuel is defined as any fuel other than oil. (Reference: Paper 3, see Table 1.)

are income, generally expressed in GDP, and fuel price levels. By comparing demand sector responses to various fuel price and income trajectories, the transport models' fuel demand elasticities are explored. Historic price and income elasticities have been reported extensively in the literature. Efficiency and service demand elasticities to fuel price are within the range of values found empirically, and very close to each other in the medium term. In 2060, the models show more diverging patterns, and elasticity values cover a broader range due to fuel substitution, increased efficiency, reduced service demand growth and feedback effects on the price shocks applied. The energy service demand projections are found to be more responsive to income level than to fuel prices, which corresponds to findings in the literature. Saturation effects of service demand over time or with increasing income are not clearly visible. Even so, the relatively small range between models has a large impact on the projected transport demand and could explain the varying transport demand growth projections, which were discussed previously.

5. Discussion and Steps Ahead

We have provided an overview of the current state of projecting future energy demand by using global models. We discussed the current methods, the results (baseline and potential for mitigation) and indicated possible improvements.

We identified future activity as a key uncertainty impacting demand sector emissions and mitigation challenges. Models projections are, relative to historically observed ranges, often quite comparable to each other. Short-term projections are similar across the various models. In the long term much larger differences can be seen mostly based on saturation dynamics and efficiency improvements. The differences across models cannot be easily traced to model type. So, while the models show an uncertain future, how this future is affected is not so well understood.

We conclude that there is a need to better understand the development of future energy demand response and the relationship with future activity levels. This is very relevant, as illustrated by the large differences in energy demand between SSP1 and SSP3. Some models use exogenous assumptions on activity levels and technology development. These can obviously limit the potential for mitigation. It must be kept in mind that models are always a simplification of reality and are therefore limited in what they can represent. This is also because they need to be transparent. Models are also limited by data and the ability to catch all relevant processes in formal equations. Therefore, using multiple scenarios, such as through the SSP framework, and the sensitivity analysis are important tools to improve our understanding of energy demand response. It also remains important to use model results in combination with other tools.

Our results are limited not only by the subset of models that we have assessed, but also by the level of depth by which the sectoral models have been studied. While we have addressed several issues for all sectors, we have studied one specific case study in detail — modeling a technology transition to electric vehicles to reflect on how demand models can be improved. Based on this work, we would finally like to briefly discuss some key areas of progress in the field of global energy demand modeling.

- **Behavior change:** Behavioral factors are difficult to quantify. Still their influence cannot be ignored. By calibrating model outcomes to historic data behavioral factors, such as technology or lifestyle preferences, behavior factors are often implicitly included. But this allows for only a limited interpretation of the effect of behavior and behavioral change in energy demand decisions. Possible next steps would be to explicitly and dynamically include empirically found relationships in the model, or assess uncertainties and possible effects through qualitative scenario design if quantification fails.
- **Physical demand:** Currently, the growth of the transport, industry and buildings sector is often directly related to exogenous economic projections. Including physical energy service demand projections, such as kilometers traveled or tonnes of steel produced, could allow us to compare energy efficiency improvements to technical bottom-up studies. The demand sector developments are also related to each other. Including physical demand projections can also be used to make cross sector relationships more explicit. Increased usage of cars, for example, would likely affect industrial sectors that provide the materials to builds cars and roads. These cross linkages between sectors relations would be relevant for further exploration. Moreover, this might enable us to better understand physical limits and possible energy demand saturation.
- **Representation of climate policy:** The more sustainable developments assumed by SSP1, such as faster or improved technology development affecting energy efficiency, or car sharing and material recycling reducing the demand for energy services, can be affected by

policy. Similarly, structural economic change, such as moving to a more service sector-oriented economy, can reduce energy requirements. However, these developments show little response to the representation of climate policy in IAMs by a carbon tax. Sector specific studies emphasize the importance of these SSP1 types of measures and indicate that they might be able to contribute more to energy demand reduction as a measure to mitigate emissions than currently accounted for under cost optimizing assumptions. Finding a way to better evaluate these types of measures in global long-term models is an important next step.

- **Barriers:** There are promising opportunities to reduce energy demand through behavioral change, sustainable developments and demand side-oriented policies. However, there are also many factors that prevent this available potential from being used. The long lifetime of energy infrastructure, institutional inertia, outdated regulations and governance structures are examples of factors that create a barrier to a possible transition. These barriers affect the energy demand sector response and are a key part of the story. Therefore, models should not only look at the available potential but also better understand and represent the barriers that need to be overcome.

- **Short-term and long-term dynamics:** To date, global models have been used primarily to show the general characteristics of mitigation strategies. After the Paris agreement, there is an increasing interest in more specific policy advice. This requires greater detail, also on the demand side. Models could follow two strategies. A group could choose to include a richer amount of technology options and/or sector details. These models would be particularly useful for analyzing short-term mitigation potentials and can be compared to sector specific studies. A second group could remain using a less detailed representation. This means that they will also be less bounded by characterization of specific technologies and be more flexible in the long term. Based on our findings we conclude that there is no preferred method, as each has their own value and can complement the other, and the type of model applied will depend on the question addressed. Both types of models are informative and it will be useful to combine and communicate the results together. Elasticities could be a way to communicate or compare results across different model types. As countries have started to implement climate policies, the translation between short-term measures and long-term trends is as relevant as ever.

Table A.1.   Main industry model characteristics.

IAM	Industry sector drivers	Industrial sub-sector breakdown	Technology	Efficiency improvements
AIM-CGE	Industrial energy demand derived through CES production function where energy is nested with industrial value added.	Iron and steel,[a] chemicals,[a] non-metallic minerals,[a] food processing, pulp and paper,[a] construction, others.	No	CES nesting structure determines the technological energy efficiency and fuel use.
DNE-21+	Material demand is related to production, consumption, import, export, population and GDP.	Iron and steel,[a] cement,[a] pulp and paper,[a] aluminium, some chemicals[a] (ethylene, propylene and ammonia).	Yes	Exogenous per technology. More efficient technologies get a larger market share in response to higher fuel prices.
GCAM	Endogenously from land use model (for fertilizer), and total GDP (for the remaining industry).	Cement,[a] nitrogenous fertilizers,[a] others.	No, only for CCS.	Technology improvement rates take into account the opportunities for improved energy efficiency, and are a scenario input assumption.
Imaclim-R	Endogenously from the equilibrium point between the supply and demand of industrial goods.	None.	No, only for CCS in cement and fertilizer.	Autonomous, and fuel price induced energy efficiency.
IMAGE	Material demand is related to economic activity and material intensity for steel and cement; energy intensity for other sectors.	Steel,[a] cement,[a] others.	Steel, cement.	Exogenous per technology more efficient technologies get a larger market share in response to higher fuel prices.
MESSAGE	Total energy demand is related to GDP and population, based on historical energy intensity trends.	Thermal and electric demand of total industry, non-energy use, cement process emissions.	No, only CCS for process carbon dioxide emissions explicitly represented.	Improvement of energy intensity depends on long-term price development. Fuel switching implies efficiency changes. No explicit representation of energy efficiency technologies.
POLES	Energy demand in industry depends on energy costs (short and long term effects) and an activity variable that is sub-sector dependent.	Iron and steel,[a] chemicals and petrochemicals,[b] non-metallic minerals,[b] others.	Boilers are described with a fixed cost, an efficiency and a life-time.	Improvement of energy intensity depends on long-term price elasticities. No explicit representation of energy efficiency technologies.
TIAM-UCL	GDP and other economic activity to derive energy demand or material demand.	Pulp and paper,[a] chemicals,[b] iron and steel,[a] non-metallic minerals,[a] others.	Yes.	Exogenous per technology more efficient technologies get a larger market share in response to higher fuel prices.

Notes: [a]Modeling physical production and energy demand of the subsector; [b]Modeling energy demand of the subsector; [c]Transformation and own energy use; [d]The process emission that can be assigned to a specific subsector.

Source: Information acquired primarily from an extensive questionnaire filled in by the model teams during the FP7 EU ADVANCE project (http://www.fp7-advance.eu/).

Policy measures	Policy impact	Material trade (industrial goods)	Stock turnover	Recycling	Energy use as feedstock	Energy use in coke oven and blast furnaces[c]	Process emissions[d]
Carbon tax or emission constraint with carbon tax.	Price mechanisms.	Yes	No	No	Only iron and steel.	Only blast furnaces.	From cement.
Carbon pricing, efficiency standards, and sectoral intensity targets.	Implementation rates of technologies and price mechanism.	Yes (exogenous scenario).	Yes	Yes	Yes	In steel sector: Yes, other sectors: No.	From cement, iron, etc.
Carbon taxes, emission constraints.	Modified fuel choices, production technologies and demands for industrial goods.	No	No	No	Yes	Yes	From cement.
Carbon/energy taxes (or energy subsidies), emissions permits.	Price mechanisms.	Yes	Yes	Yes, but not explicitly.	No	No	No
Carbon tax, prescribing certain efficient technologies.	A dynamic response to changed technology costs (incl. fuel price) or prescribed technology mix.	Yes, only for cement and steel.	Yes	Yes	Yes	Yes	From cement.
GHG and energy pricing, GHG emission cap, permits trading, fuel subsidies, capacity, production and share target regulations.	Price mechanisms and model constraints.	No	No	No	Yes	In steel sector: yes, other sectors: No.	From cement.
Taxation policy on energy fuels, which includes carbon pricing.	Price mechanism.	Yes	(only for boilers)	No	Yes	Only own energy use in blast furnaces.	From cement.
Carbon tax/cap, permit trading, technology subsidy, efficiency requirement.	Price mechanisms and model constraints.	Yes, but not explicitly modeled.	Yes	No recycling.	Yes	Yes	No

Table A.2.   Drivers of energy demand in the transport sector of 11 IAMs.

	TIAM-UCL	IMAGE	Imaclim-R
System boundaries	The fuel mix is determined endogenously. Indirect fuel use from manufacturing, upstream energy and emissions are calculated but not tied to transport.	The model determines the fuel use, which is linked to the TIMER model, hence all emissions from fuels are considered. Embodied emissions of vehicles are included in the industry sector.	As a CGE model all GHG-emitting and energy producing/consuming sectors are included. This implies that indirect energy use and emissions from fuel production and vehicle manufacture are included, but in the energy transformation and industry sectors.
Relationship drivers and demand	GDP, population, and GDPP drive the transport demand, where energy service demand grows slower than the underlying driver. The demand is influenced through a linear relationship with the drivers. Each transport demand in each region has its own relationship driver and demand coupling factor.	GDP, IVA (for freight) population, fuel price, non-energy price, load factor, mode preferences, energy efficiency, mode speed drive service demand per mode, on the basis of Travel money budget (TMB) and Travel time budget (TTB) formulation. A fleet module determines fleet composition within each mode, affecting mode cost, energy efficiency and fuel type for each mode.	The mobility demand and modal split result endogenously from a household's utility maximization under constraints of revenues and time spent in transport. Each mode is characterized by a price and a speed. The price of cars mobility depends on fuel prices and the cost of car ownership, while other modes by the intermediate consumption shares and prices within the general equilibrium framework. When infrastructure use reaches congestion, the marginal speed of the mode decreases, which limits its use.

	REMIND	GCAM	AIM-CGE
System boundaries	Input of final energy in different forms is required together with investments and operation and maintenance payments into the distribution infrastructure as well as into the vehicle stock. Material needs and embodied energy are not considered.	The full fuel cycle of each fuel is represented. This includes biomass from an agriculture and land use model. No other upstream inputs to the sector are considered (e.g., vehicle manufacturing, roads).	Indirect energy use is treated in the energy transformation sector.
Relationship drivers and demand	GDP growth, the autonomous efficiency improvements, the elasticities of substitution between capital and energy and between stationary and transport energy forms. Mobility from the different modes is input to a CES function, the output of which is combined with stationary energy in a CES function to generate a generalized energy good, which is combined with labor and capital in the main production function for GDP.	GDP, population, and services prices, derived from vehicle speeds and vehicle levelized average operating costs. GDP sets the scale of the demand, and determines the wage rate, which determines the opportunity cost of each travel mode. In this way, increases in GDP will increase the per capita demand for travel, and shift this demand towards the fastest modes.	Transport intermediate inputs and final demand. Passenger transport is determined by GDP with elasticity. Freight transport is determined by all industrial sectors inputs. They are formulated as multiplying input coefficient.

*The MESSAGE transport module used in this study is a simpler version than used in other papers (e.g., McCollum *et al.*, 2017). Specifically, this version is MESSAGE-Transport V.5; yet, for the purposes of this paper, the model did not make any explicit assumptions about heterogeneous behavioral features among consumers.

MESSAGE*	POLES
All GHG-emitting and energy producing/consuming sectors are included. This implies that indirect energy use and emissions from fuel production and vehicle manufacture are included, but the latter is not represented by a direct linkage.	The transportation sector covers the transport of goods and passengers. Transport of energy and associated losses, which are accounted for in the own energy uses of the energy sector.
Fuel prices, vehicle costs, GDP, population, vehicle speeds, vehicle occupancy rates, passenger vehicles per capita, annual distance traveled per vehicle, etc. Travel money budget, travel time budget, income, travel prices and travel speed determine service demand for the different modes (mode choice). The optimization framework determines the fleet composition within each mode. Freight service demand is driven by population, GDP and price elasticity.	Passengers: • Cars: income increase the number of cars per capita, fuel price affects the yearly mileage. • Rail and buses: income increase the mobility, fuel price increase modal shift from cars to public transport. • Goods: GDP growth affects the mobility per mode.

DNE21+	GEM-E3	WITCH
Indirect energy use is not included. For example, emissions from the car manufacturing process are classified into the industrial sector.	All GHG-emitting and energy producing/ consuming sectors are represented explicitly in the model.	LDV and road freight are explicitly modeled, while other modes are embedded within a non-electric sector. Aspects such as infrastructure and the vehicle manufacturing are incorporated in the overall GDP and representation of final goods.
Scenarios on service demand of road transportations are developed for passenger cars and buses separately based on per capita GDP and historical trends. As for road freight transport scenarios of cargo trucks, overall cargo service per capita is estimated by the GDP size, under assumption of modal shifts.	The mobility demand and its modal split result endogenously from a household's utility maximization under constraints of income and firms under maximization revenues. Each mode is characterized by a price. The price of car mobility depends on operational cost and the purchased cost. The price of other modes is determined in the general equilibrium framework by the intermediate consumption shares and prices.	A linear Leontief function combines energy, O&M, vehicle capital and carbon costs to select the optimal mix of vehicle types. Vehicle ownership is a main driver, which is set via a calibration based upon GDP growth. Exogenous efficiency improvements are implemented within the model.

Table A.3. Transport sector technologies and final energy carriers.

	TIAM-UCL[23]	IMAGE[24]	Imaclim-R[25]
Modes and vehicle types	Passenger: 7 modes (two wheel, three wheel international aviation, domestic aviation, road auto, road bus, rail), Freight: 7 modes (light, commercial, medium, heavy truck, rail, domestic navigation, international navigation), and hundreds of technologies.	Passenger: 7 modes (walk, bicycle, bus, train, car, high speed train and airplane), 6 freight modes (national ship freight, international ship freight, medium truck, heavy truck, rail freight, air freight). Tens of technologies per mode.	Passenger: 4 modes (non motorized, personal vehicles, airplane, other) and 3 freight (trucks & freight rail, airplane, shipping). Technologies: ICE, efficient ICE, hybrid, plug-in hybrid and electric.
Final energy carriers	Diesel, Gasoline, Ethanol, Electricity, LPG, Methanol, Natural Gas, Hydrogen, Fischer Tropsch biofuels.	The transport model only considers the secondary energy carriers: Hydrogen, Gas, Electricity, Oil, and Biofuel.	Liquid fuels from oil, Synthetic liquid fuels from other fossils Liquid fuels from biomass, and electricity.
Energy consumption of vehicles	Share estimates split fuel consumption between road modes and rail modes. The model invests in technologies in order to satisfy the energy service demands in order to maximize consumer and producer surplus. Final energy consumption is endogenous to the model solution.	Different vehicle types with different energy efficiency's compete against each other (based on the multinomial logit), which allows for a change of energy efficiency of the mode.	For personal vehicles: explicit technologies with an efficiency characteristic and leaning on the cost. For other modes: efficiency improvement triggered by fuel prices.
Determinants technology costs and shares	Investment costs, O&M costs, fixed costs — are based on exogenous assumptions and change over time in response to an exogenous learning curve. Vehicle market share is outcome of the model solution.	Net present costs based on literature, decreasing exogenously in time. We assumed that the technology costs is a global variable, as the technologies tend to be traded worldwide. Vehicle share is based on a multinomial logit.	All technology characteristics are fixed in time, except costs that endogenously decrease with a learning rate. Vehicle market share is based on a logit function.
Distribution between transport modes	Distribution is assumed exogenously, but the split between modes may slightly change due to responses to own price elasticities.	Time and costs are considered. Cost are weighted relative to time with a time-weight factor. The time-weight factor is determined by the travel money and travel time budget.	Households utility maximization under both constraints of revenues and time.

	REMIND[28]	GCAM[29]	DNE21+[30]
Modes and vehicle types	Passenger: 4 modes, Freight: 1 mode. For passenger transport: LDV, Aviation, Bus and Electric Trains. One generic freight transport.	Passenger: 10 modes. Freight: 4 modes. Off-road vehicles, mining, or agriculture are not part of the transportation sector, except for China and India. ICE, electric, hybrid, fuel cell and compressed natural gas for bus/passenger. For other modes one or two technology options.	Road transportation: 5 modes. The other sub-sectors are generated in a top-down manner. Technologies: ICEs, ICE efficient, HEV, PHEV, electric, fuel-cell.
Final energy carriers	Liquids: Coal, Gas, Oil or Biomass (only second-generation with CCS for Coal and Biomass). Electricity (only LDV). Hydrogen (only LDV) (Coal, Gas or Biomass, all combined with CCS).	Liquid fuels (includes fuels derived from oil, coal, gas and biomass). Electricity Natural gas (mostly natural gas; also includes biogas and coal gas). Hydrogen (from many fuels). Coal (for rail in China).	Gasoline, Diesel, Bioethanol and Biodiesel, CNG, Electricity, Hydrogen from coal, gas biomass and electricity, plus CTL (coal to liquid) and CTG (coal to gas).
Energy consumption of vehicles	The general efficiency of one transport mode improves exogenously over time in the CES function.	The energy quantity is derived from the average vehicle intensity and the load factor. The energy intensity of each technology is assumed to change over time exogenously. Endogenous changes of energy intensity are due to (a) switching from ICE to hybrid vehicles, (b) switching from smaller to larger vehicles, (c) modal shifting, or (d) switching to fuels with lower end-use energy intensity.	Energy consumption is determined based on the exogenous scenarios on service demand of road transportations in combination with technology (fuel efficiency of vehicles, costs and implicit discount rate) choice.
Determinants technology costs and shares	Efficiency, lifetime, investment cost, fixed O&M. Investment cost for battery electric and fuel cell vehicles decrease endogenously following a global learning rate towards a given floor cost. The distribution of LDV vehicles follow cost optimization with different non-linear constraints.	Capital costs are amortized over an exogenous lifetime, assuming a 10% discount rate. Non-fuel operating costs include insurance, registration, taxes and fees, and standard O&M expenses. These can decrease exogenously for immature technologies such as electric cars or hybrid vehicles. Vehicle market share is based on a logit function.	Fuel efficiency of vehicles and costs are assumed to be improved exogenously. Lifetime does not change over time. Market shares are based on least cost optimization.
Distribution between transport modes	The distribution between LDV and other modes is determined via the CES production function, driven by the elasticity of substitution (1.5) and the evolution of the efficiency parameters.	The modes compete using a logit share formulation, where the costs include both the vehicle cost and the time value cost. The time value cost is derived as the wage rate divided by the average transit speed, and modified by an exogenous time-value multiplier that is generally close to 1.	Travel demand is exogenously given for each mode. Modal shift is not endogenously evaluated.

* The MESSAGE transport module used in this study is a simpler version than used in other papers (e.g., McCollum *et al.*, 2017). Specifically, this version is MESSAGE-Transport V.5; yet, for the purposes of this paper, the model did not make any explicit assumptions about heterogeneous behavioral features among consumers.

MESSAGE*[26]	POLES[27]
passenger modes and 1 freight mode. Other modes are not explicitly modeled but their energy use is accounted for via an exogenous energy demand trajectory. Tens of technologies options per mode.	Passengers: 7 modes (cars, motorbikes, bus, rail, air). Goods: 5 modes (heavy vehicles, light vehicles, rail, other (inland water), maritime). Technologies: ICE, plugin hybrid-electric, battery electric, fuel cell.
..ll fuels from the MESSAGE energy systems model are considered in the transport module.	Oil products, Biofuels (energy crops and cellulosic feedstocks), Gas, Coal (for rail), Electricity and Hydrogen.
)ifferent vehicle types with different energy efficiencies compete against each other, which allows for an average change of energy efficiency of the mode over time. The techno-economic parameters for each technology are exogenously assumed.	Unit consumption depends on: • Price: long term elasticity to account for investment and short term to account for behavior. • Income for behavior, to control the spending on fuel for transportation (maximum "budgetary coefficient").
'he techno-economic parameters are exogenously assumed and change over time. There is also regional differentiation for certain technologies and parameter assumptions. Market shares are based on least cost optimization.	Road vehicles: Efficiency, lifetime, investment cost, fixed and variable O&M. These parameters change overtime exogenously. Vehicle competition based total user cost and infrastructure possible development.
'ime and costs are considered. Costs are weighted relative to time with a time-weight factor. The time-weight factor is determined by the travel money and travel time budget.	The different modes are mostly disconnected, limited by: differentiated elasticities to fuel prices and saturation effects (e.g., maximum number of cars per capita, maximum air related mobility).

GEME3[31]	AIM-CGE[32]	WITCH[33]
assenger: 5 modes (Passenger Cars, LDV/Bus, Aviation, rail and inland navigation). Freight: 3 modes (LDV/heavy trucks, rail, inland navigation). Technologies: pure conventional, hybrid, plugin hybrid-electric, battery electric, biofuels.	Five passenger modes (bus, train, car (incl. 2- and 3-wheelers), train, airplane). Freight: 6 modes (national ship freight, international ship freight, medium truck, heavy truck, rail freight, air freight). Aggregated technology.	Two modes. Road passenger and freight, both featuring four vehicle types: ICE, hybrid, plug-in hybrid and battery electric.
Road: Oil, Electricity, Gas, Bio-gasoline and Biodiesel (traditional and second generation). Rail: Coal, Oil, Biodiesel and electricity. Airplane: Oil, Biodiesel. Ship: Oil, Biodiesel.	Road: Oil, electricity, and biofuel (bus can use gas), Railway: electricity and coal. Ship: oil, biofuel and coal. Airplane: oil and biofuel.	Liquids can come from Oil or Biomass (traditional or second-generation). Electricity can come from coal (possibly with CCS), gas, oil, biomass (possibly with CCS), wind, PV, CSP, hydro or nuclear
)ifferent passenger car types with different energy efficiencies compete against each other based on Weibull. The efficiency of other transport modes improves exogenously over time in the CES function	Multiplying coefficient. Fuel efficiency improvement is considered.	The efficiency of LDV and road freight transport modes improves exogenously over time based on selected efficiency improvement targets or selected forecasts.
)apital costs are amortized over an exogenous lifetime, assuming a 12.5% discount rate. Non-fuel operating costs include insurance, registration, taxes and fees, and standard O&M expenses.)apital cost decrease endogenously for immature technologies such as electric cars or plugin hybrid vehicles assuming a global learning rate towards a given floor cost. Vehicle market share is based on Weibull function.	Not explicitly determined.	Efficiency, lifetime, investment cost, fixed O&M. Investment cost for battery electric vehicles decreases following a global learning rate as a consequence of endogenously modeled investments in R&D. The distribution of LDV and road freight vehicles follows cost optimization with different non-linear constraints.
The different type of passenger cars compete using a Weibull share formulation, where the costs includes both operational cost and purchase cost. The distribution between LDV and other modes is determined via the CES production function, driven by relative prices and the evolution of the efficiency parameters.	—	The distribution between modes is fixed and determined via separate demand calculations.

References

1. V. Krey, "Global energy-climate scenarios and models: a review", *Wiley Interdisciplinary Reviews: Energy and Environment* **3** (2014) 363–383.
2. E. Stehfest, D.P. van Vuuren, L. Bouwman and T. Kram, *Integrated Assessment of Global Environmental Change with IMAGE 3.0: Model Description and Policy Applications* (Netherlands Environmental Assessment Agency (PBL), 2014).
3. D.P. van Vuuren, M.T.J. Kok, B. Girod, P.L. Lucas and B. de Vries, "Scenarios in global environmental assessments: key characteristics and lessons for future use", *Global Environment Change* **22** (2012) 884–895.
4. H.G. Huntington, J.P. Weyant and J.L. Sweeney, "Modeling for insights, not numbers: the experiences of the energy modeling forum", *Omega* **10** (1982) 449–462.
5. M.B.A. Van Asselt and J. Rotmans, "Uncertainty in Integrated Assessment modelling. From positivism to pluralism", *Climate Change* **54** (2002) 75–105.
6. R.H. Moss *et al.*, "The next generation of scenarios for climate change research and assessment", *Nature* **463** (2010) 747–756.
7. B.C. O'Neill *et al.*, "A new scenario framework for climate change research: the concept of shared socioeconomic pathways", *Climate Change* **122** (2014) 387–400.
8. E. Kriegler *et al.*, "A short note on integrated assessment modeling approaches: rejoinder to the review of making or breaking climate targets: the AMPERE study on staged accession scenarios for climate policy", *Technological Forecast and Social Change* **99** (2015) 273–276.
9. E. Kriegler *et al.*, "Socio-economic scenario development for climate change analysis", 2010.
10. T. Morita, N. Nakićenović and J. Robinson, "Overview of mitigation scenarios for global climate stabilization based on new IPCC emission scenarios (SRES)", *Environmental Economics and Policy Studies* **3** (2000) 65–88.
11. P.W.F. Van Notten, J. Rotmans, M.B.A. Van Asselt and D.S. Rothman, "An updated scenario typology", *Futures* **35** (2003) 423–443.
12. M. Sugiyama *et al.*, "Energy efficiency potentials for global climate change mitigation", *Climate Change* **123** (2014) 397–411.
13. O.Y. Edelenbosch *et al.*, "Comparing projections of industrial energy demand and greenhouse gas emissions in long-term energy models", *Energy* **122** (2017) 701–710.
14. O.Y. Edelenbosch *et al.*, "Decomposing passenger transport futures: comparing results of global integrated assessment models", *Transportation Research Part D, Transport and Environment* **55** (2017) 281–293.
15. O.Y. Edelenbosch *et al.*, "Transport fuel demand responses to fuel price and income projections: comparison of integrated assessment models", *Transportation Research Part D, Transport and Environment* **55** (2017) 310–327.
16. O.Y. Edelenbosch, D.P. van Vuuren, K. Blok, K. Calvin and S. Fujimori, "Mitigating energy demand sector emissions: the integrated modelling perspective", *Applied Energy*, 2018.

17. B.W. Ang, "Decomposition analysis for policymaking in energy: which is the preferred method?" *Energy Policy* **32** (2004) 1131–1139.

18. UN Environment, *Emissions Gap Report: A UNEP Synthesis Report*, 2017.

19. K. Riahi *et al.*, "The Shared Socioeconomic Pathways and their energy, land use, and greenhouse gas emissions implications: an overview", *Global Environmental Change* **42** (2017) 153–168.

20. K. Kermeli, W.H.J. Graus and E. Worrell, "Energy efficiency improvement potentials and a low energy demand scenario for the global industrial sector", *Energy Efficiency* **7** (2014) 987–1011.

21. IEA, *World Energy Outlook 2016*, 2016.

22. F. Creutzig *et al.*, "Transport: a roadblock to climate change mitigation?" *Science* **350** (2015) 911–912.

23. G. Anandarajah, S. Pye, W. Usher, F. Kesicki and C. McGlade, "TIAM-UCL global model documentation", *University College London*, 2011.

24. B. Girod, D.P. van Vuuren and S. Deetman, "Global travel within the 2C climate target", *Energy Policy* **45** (2012) 152–166.

25. H.-D. Waisman, C. Guivarch and F. Lecocq, "The transportation sector and low-carbon growth pathways: modelling urban, infrastructure, and spatial determinants of mobility", *Climate Policy* **13** (2013) 106–129.

26. K. Riahi *et al.*, "Energy pathways for sustainable development", 2012.

27. B. Girod *et al.*, "Climate impact of transportation: a model comparison", *Climate Change* **118** (2013) 595–608.

28. G. Luderer *et al.*, "The economics of decarbonizing the energy system: results and insights from the RECIPE model intercomparison", *Climate Change* **114** (2012) 9–37.

29. P. Kyle and S.H. Kim, "Long-term implications of alternative light-duty vehicle technologies for global greenhouse gas emissions and primary energy demands", *Energy Policy* **39** (2011) 3012–3024.

30. F. Sano, K. Wada, K. Akimoto and J. Oda, "Assessments of GHG emission reduction scenarios of different levels and different short-term pledges through macro-and sectoral decomposition analyses", *Technological Forecast and Social Change* **90** (2015) 153–165.

31. P. Karkatsoulis, P. Siskos, L. Paroussos and P. Capros, "Simulating deep CO_2 emission reduction in transport in a general equilibrium framework: the GEM-E3T model", *Transportation Research Part D, Transport and Environment* **55** (2017) 343–358.

32. S. Fujimori, T. Masui and Y. Matsuoka, "Development of a global computable general equilibrium model coupled with detailed energy end-use technology", *Applied Energy* **128** (2014) 296–306.

33. V. Bosetti and T. Longden, "Light duty vehicle transportation and global climate policy: the importance of electric drive vehicles", *Energy Policy* **58** (2013) 209–219.

Index

About the Editor

Henry Kelly is a Non-resident Senior Fellow at Boston University's Center for Sustainable Energy. He has held a number of positions in the federal government including the US Department of Energy (Principle Deputy Assistant Secretary and later acting Assistant Secretary for Energy Efficiency and Renewable Energy), the White House Office of Science and Technology Policy the US Arms Control and Disarmament Agency, and the Congressional Office of Technology Assessment. He has also served as the Assistant Director of the Solar Energy Research Institute (now the National Renewable Energy Laboratory) and President of the Federation of American Scientists. He is an elected fellow of the AAAS and the American Physical Society and is a member of the Advisory Board for the Energy and Environment Directorate of the Pacific Northwest National Laboratory. He has a Ph.D. in physics from Harvard University and a B.A. in physics from Cornell University.

https://doi.org/10.1142/9789811217869_bmatter

About the Contributors

Peter Alstone is a Faculty Scientist at the Schatz Center, Assistant Professor of Environmental Resources Engineering at Humboldt State University, and Faculty Affiliate researcher at Lawrence Berkeley National Laboratory. Peter's research approach is interdisciplinary, with a thematic focus on technology and policy for distributed energy systems. His recent work includes demand response potential modeling for California, market transformation programs for global access to reliable electricity, and systems integration for advanced microgrids.

Tia Benson Tolle graduated from the University of Washington with a Bachelor of Science degree in Mechanical Engineering. She also earned her Master of Science and Doctorate of Philosophy degrees in Materials Science and Engineering from the University of Dayton. In addition, she holds a Master's Certificate in Leadership and Executive Development from the University of Dayton and completed the Air Force Senior Leadership Development Course and Air War College Senior Leader Course from the Air University, Maxwell Air Force Base.

Tia joined NASA's Johnson Space Center 1983 as a co-op student and in 1986 as a Flight Crew Instructor in the Space Shuttle Flight Training Division, Mission Operations Directorate.

In 1989, she joined the Flight Dynamics Laboratory, Wright Laboratory, as a Composite Structures Program Manager in the Advanced Composites Advanced Development Program Office.

Tia joined the Materials and Manufacturing Directorate, Air Force Research Laboratory, in 1992. She held several positions, including Chief of the Structural Materials Branch and lead of the Composites Core Technology Area. During this period, she was responsible for materials for satellites, airframes, engines, and launch vehicles spanning basic research to advanced development and transition. In 2007, she was assigned as the Technology Director of the Nonmetallic Materials Division. She was also USAF Deputy to the DoD Reliance Community of Interest for Materials and Processes, coordinating research interests across Defense Science and Technology.

In 2012, Tia joined Boeing as the Director of Advanced Materials, Product Development, at Boeing Commercial Aircraft. In this position, she is accountable for the integrated materials portfolio for Boeing's commercial airplane products.

Tia is a Fellow at the Society for the Advancement of Material and Process Engineering (SAMPE), an International Past President of SAMPE, and was President of the Materials Research Society (2014). She is the BCA Executive Sponsor for Boeing's SAMPE External Technical Affiliation as well as Edmonds Community College. She serves on Iowa State University's Aerospace Engineering Department's Industry Advisory Council, the University of Washington Materials Science & Engineering Department's Industry Advisory Council, and is a Trustee of Edmonds Community College.

Gaudy M. Bezos-O'Connor is a Project Manager for the Electrified Powertrain Flight Demonstration (EPFD) Project within NASA Aeronautics Research Mission Directorate's Integrated Aviation Systems Program. The EPFD Project is a cost-shared, public-private partnership focused on maturing MW-class electrified powertrain technology to accelerate product introduction into the U.S. commercial transport fleet. The flight demonstrations planned in the 2023–2024 timeframe will retire powertrain and propulsion airframe integration technical barriers to entry and address associated certification and standards gaps. She recently completed a detail as a Program Engineer in the FAA's Continuous Lower Energy, Emissions and Noise (CLEEN) Program which is focused on advancing subsonic and supersonic technologies in partnership with NASA, industry and academia in the achievement of green aviation emissions, community noise, and fuel efficiency goals. Prior to the FAA detail, she served as the Deputy Project Manager for the Advanced Air Transport Project under NASA's Advanced Air Vehicles Program as well as the Deputy Project Manager for NASA's Environmentally Responsible Aviation Project under the Integrated Aviation Systems Program. Under her leadership, the ERA and AATT Projects in partnership with the aviation industry and partner federal labs have matured green aviation airframe and propulsion technologies targeted for new commercial subsonic transports in the 2025–2035 timeframe. Ms. Bezos-O'Connor's career at NASA began at Langley Research Center's Aeronautics Directorate where she conducted research in the fields of subsonic and supersonic aircraft performance including the FAA-NASA Windshear Hazards Program, F16XL Supersonic Laminar Flow Control Flight Experiment and the High Speed Civil Transport Program. Ms. Bezos-O'Connor is an AIAA Associate Fellow and serves on the AIAA Green Engineering Integration Committee. She is the recipient of NASA's Exceptional Engineering Achievement Medal. Her team awards include the NASA Silver Achievement Medal and the Blue Marble Environment and Energy Award, the Royal Aeronautical Society's 2017 Team Specialist Gold Award, and Aviation Week's 2016 Laureate Technology Award, all for the ERA Project. Ms. Bezos-O'Connor received a Bachelor of Science degree in Aeronautical Engineering from Rensselaer Polytechnic Institute and a Master of Engineering in Engineering Management from Old Dominion University.

Adam Cohen is a transportation mobility futures researcher at the Transportation Sustainability Research Center at the University of California, Berkeley. Since joining the group in 2004, his research has focused on innovative urban mobility strategies, including urban air mobility, automated vehicles, shared mobility, smart cities technologies, smartphone apps, and other emerging technologies. He has co-authored numerous articles and reports in peer-reviewed journals and conference proceedings. Previously, Cohen worked for the Information Technology and Telecommunications Laboratory (ITTL) at the Georgia Tech Research Institute (GTRI). His academic background is in city and regional planning and international affairs.

Doug Christensen is an Associate Technical Fellow and the Technical Leader for the Boeing Company's ecoDemonstrator technology demonstration vehicles. He has overall responsibilities for the flight demonstration aircraft acquisition, technology demonstrations, and external ecoDemonstrator program activities.

Prior to his ecoDemonstrator role, Doug was the technology leader for NLF wing and winglet design/build, as well as the CO_2 and Emissions Strategy Leader within Boeing Commercial Airplanes Product Development. He also led the development and maintenance for all of BCA's environmental technology development.

Doug's background is in configuration design and analysis and he spent 15 years as an Aerodynamics Configuration Engineer in Product Development and as an Airplane Configurator in Configuration and Engineering Analysis.

Christensen also worked for Northwest Airlines as a Senior Performance Engineer for 5 years where he was involved in the integration of new aircraft into their fleet and developed numerous fuel burn reduction processes.

He joined Boeing in 1987 after graduation from Iowa State University where he received a Bachelor of Science degree in Aerospace Engineering.

Stéphane de la rue du Can is a Program Manager and the Assistant Group Leader of the International Energy Studies Group. She joined Berkeley Lab in 2003, after working 3 years at the International Energy Agency in Paris. Stephane's main contributions at Berkeley Lab have been in improving the scientific understanding of end-use energy consumption in developing countries, CO_2 emissions accounting methodologies, modeling future growth, assessing energy savings, developing emissions mitigation pathways, conducting cost benefit analysis, and promoting innovative energy efficiency policy. Stéphane has been a member of the Intergovernmental Panel on Climate Change (IPCC) since 2003 and was a contributing author to the Fourth Report which shared the Nobel Prize with Al Gore in 2008.

Rick Diamond, Guest Scientist at LBNL, worked for over 40 years at LBNL on issues of energy, behavior, and buildings. He taught previously at the Graduate School of Design, Harvard University, and the Department of Architecture, UC Berkeley. He is currently interested in resilient buildings and communities, and forest fires and health.

Oreane Edelenbosch works at the PBL Netherlands Environmental Assessment Agency, focussing in particular on the contribution of energy efficiency and demand changes in buildings, industry and transport to global climate scenarios. She has held positions at Utrecht University, where she obtained her PhD, Polytechnic University of Milan, as a post-doc researcher where her work focused on modelling the impacts of behavior and heterogeneity on energy consumption. She is an associated researcher at RFF-CMCC European Institute on Economics and the Environment. She contributed to the UNEP Emissions Gap Report 2017 and IPCC special report on 1.5°C global warming.

Jessica Granderson is a Staff Scientist and the Deputy of Research Programs for the Building Technology and Urban Systems Division at the Lawrence Berkeley National Laboratory. She is a member of the Whole Building Systems Department. Dr. Granderson holds a Ph.D. in Mechanical Engineering from UC Berkeley, and an A.B. in Mechanical Engineering from Harvard University. Her research focuses on building energy performance monitoring and diagnostics, advanced measurement and verification, and intelligent lighting controls. She is the recipient of the 2015 Clean Energy Education and Empowerment (C3E) Award for Leadership in Research.

Tianzhen Hong is a Staff Scientist at the Building Technology and Urban Systems Division of Lawrence Berkeley National Laboratory. He is an IBPSA Fellow and leads the Urban Systems Group and a research team working on building energy modeling and simulation, data analytics, machine learning, computational tools, and policy to advance design, operation, and retrofit of buildings and urban systems.

Paul Mathew is a Staff Scientist and Department Head of Whole Building Systems at LBNL, where he conducts applied research and market transformation activities on energy use in buildings. His current work is focused on integrated building systems, energy epidemiology, benchmarking tools, and energy-related risk analysis. Prior to joining LBNL, he worked at Enron Energy Services and the Center for Building Performance at Carnegie Mellon University. He has authored over 100 technical papers, articles, and reports. He received a U.S. presidential award for federal energy efficiency in 2007. He has a Bachelor's degree in Architecture, and a Ph.D. in Building Performance and Diagnostics from Carnegie Mellon University.

Alan Meier is a Senior Scientist at LBNL, where he leads the Electronics, Lighting, and Networks group. Meier was instrumental in identifying the problem of standby energy use in appliances and equipment, and then devising technologies to reduce it. Meier's research focuses on using enhanced communication technologies and networks to economically reduce energy consumption. Recently, he developed techniques to track power outages with Internet-connected thermostats. Meier is also an adjunct professor of Environmental Science and Policy at the University of California, Davis.

Erik Page is a leader in the field of energy-efficient lighting with broad expertise across many areas including LEDs, fluorescent and compact fluorescent systems, lighting controls, luminaires, and photometry. Erik spent ten years as a researcher and lab manager in the Lighting Fixtures Laboratory at the Lawrence Berkeley National Laboratory before leaving to help establish the California Lighting Technology Center, where he served as the Director of Engineering for four years. In 2008, Erik established Erik Page & Associates, Inc., a small consulting firm based in Fairfax, CA, providing clients with technically robust support for various projects related to domestic (USA) and international energy-efficient lighting projects. He has authored dozens of publications and holds twelve US Patents.

David Paisley is a Technical Fellow in Boeing Commercial Airplanes Product Strategy and Future Airplane Development.

Dr. Paisley is the Team Leader for Airplane Configuration and Product Evaluation organization where he leads a team that that evaluates technologies, features, and concepts for application to future and derivative airplane products.

He previously worked on the environmental impact of commercial aviation, particularly in developing environmental requirements for new airplane concepts after spending a considerable amount of time working in Product Development on new commercial airplane design concepts and methods. During this time, he co-developed and co-authored the Boeing Commercial Airplanes Environmental Performance Outlook, (now an annual publication) illustrating an ~ 40-year forecast of the global environmental landscape used to anticipate, inform, and evolve to meet business needs.

Prior to that, Dr. Paisley spent several years at Boeing Helicopters in Philadelphia working on the V-22 and advanced tiltrotor design concepts following a short spell at Canadair (now Bombardier) in Montreal developing innovative rotorcraft products in the Surveillance Systems division. Dave began his career at British Aircraft Corporation (now BAe Systems) at Warton in the United Kingdom, working on advanced combat aircraft, including Jaguar and Tornado.

Dr. Paisley holds a B.Sc. degree in Aeronautical Engineering from the University of Manchester (UK), a Ph.D. in Aeronautical Engineering from Cranfield University (UK), and an M.S. in Management from Antioch University Seattle.

He is an Associate Fellow of AIAA and a member of SAE, and the SAE Environment Committee.

Mary Ann Piette is a Senior Scientist, Director of the Building Technology and Urban Systems Division, and Senior Science Advisor to the Associate Lab Director of Energy Technologies Area at LBNL. She oversees LBNL's building energy research activities with the US Department of Energy and over 25 other R&D sponsors. She is also the Director of the Demand Response Research Center. Mary Ann develops and evaluates new technology and building components, controls, operations, simulation, whole building and electric load shape analysis and behavior. She has an M.S. in Mechanical Engineering from UC Berkeley and a Licentiate in Building Services Engineering from the Chalmers University of Technology in Sweden.

Tim Rahmes is an Associate Technical Fellow in the Flight Sciences Engineering group for Boeing Commercial Airplanes.

He currently leads Boeing's atmospheric observation efforts to provide customers with cross-model (777, 787, 737 MAX, 777X, etc.) weather and turbulence observations and flight efficiency applications.

Through his career at Boeing, Tim has been the principal engineer for recent flight test research programs, and has been the principle engineer for several projects on the Boeing ecoDemonstrator Program. He has worked on weather, emissions, atmospheric sensors, alternative fuels, avionics, datalink, air traffic management, and vehicle health management applications.

After starting his career as a US Naval Flight officer, he has held positions in industry such as a systems and software engineer as well as a research scientist at South Pole Station.

Tim has a Bachelor of Science in Aeronautical Engineering from Rensselaer Polytechnic Institute, a Master of Science in Atmospheric Science from University of Illinois at Urbana-Champaign, and a Master of Business Administration from the University of Washington.

He is also a member of SAE — SAE International, EUROCAE — European Organization for Civil Aviation Equipment, AIAA — American Institute of Aeronautics and Astronautics, and RTCA. Tim has also been a co-chair for RTCA SC-206 subgroups for data link weather and aeronautical information standards.

Stephen Selkowitz is a retired LBNL scientist and an internationally recognized expert on window technologies, façade systems, and daylighting. He is a frequently invited speaker to industrial and professional groups on many aspects of building technologies and commercial building energy efficiency, and is the author/co-author of over 170 publications and 3 books, and holds 2 patents. He is a past member of the Board of Directors of the National Fenestration Rating Council and is currently on the advisory board of a number of efficiency initiatives such as the Green Lights Daylighting Program in New York City and the Zero Emissions Building program in Norway. Selkowitz holds a B.A. in Physics from Harvard College and an M.F.A. in Environmental Design from California Institute of the Arts.

Susan Shaheen is a pioneer and thought leader in emerging mobility strategies. She is a Professor in the Department of Civil and Environmental Engineering and Co-Director of the Transportation Sustainability Research Center (TSRC) at the University of California, Berkeley. She has authored 75 journal articles, over 125 reports and proceedings articles, 18 book chapters, and co-edited two books. In May 2019, she received the most influential paper award from World Conference on Transportation Research. She was named "Faculty of the Year" by the Institute of Transportation Studies, Berkeley, in December 2018 and received the 2017 Roy W. Crum award from the Transportation Research Board (TRB) for her distinguished achievements in transportation research. In May 2016, she was named one of the top 10 academic thought leaders in transportation by the Eno Transportation Foundation. She is Vice Chair of the TRB Executive Committee. She was the founding chair of the subcommittee for Shared-Use Vehicle Public Transport Systems of the TRB. She is a member of the Mobile Source Technical Review Subcommittee to the U.S. Environmental Protection Agency's Clean Air Act Advisory Committee.

Al Sipe is the Chief Architect for Boeing Operational Efficiency. In this role, he leads a team responsible for developing and/or advancing technology solutions that improve the efficiency of commercial aircraft, leveraging existing airplane technology, accelerating new capabilities to operations and new airplane requirements.

Al and his team are also responsible for international advocacy and alignment of air traffic management in support of global CO_2 fuel consumption, emissions, and noise reduction and are active members in CANSO, ATAG, CAEP WG-1/2/3, and the World ATM Congress.

Some of the key technologies he and his team have successfully implemented are in flight data trajectory optimization, ADS-B Out/In acceleration, RNP "established", GLS/3GMMR, FAA datalink, sustainable aviation fuels, digital iPads, and optimizing flight paths and airport noise among others.

Al's background is in Air Traffic-related airplane systems. In his previous position, he was Lead Systems Engineer for Boeing Operational Efficiency in Boeing Research and Development.

Previous positions include programs such as the Boeing 777, B-2 Bomber, and the Airborne Laser Programs.

Al received a Bachelor of Science, Electrical Engineer from Montana State University, and holds an Instrument Rated Private Pilot license in land and seaplanes.

Kaiyu Sun is a Senior Scientific Engineering Associate in the Building Technology and Urban Systems Division at Lawrence Berkeley National Laboratory. Her current work focuses on building energy modeling and simulation tool development, occupant behavior research, and building resilience integrated with energy efficiency.

Detlef van Vuuren (1970) is professor in Integrated Assessment of Global Environmental Change at the Faculty of Geosciences, Utrecht University, and senior researcher at the PBL Netherlands Environmental Assessment Agency, leading the IMAGE integrated assessment modeling team. He has published more than 300 articles in refereed journals including Nature, Science, Nature Climate Change, Nature Energy, Nature Geosciences, PNAS, and Environmental Research Letters. He is listed among the most highly cited researchers worldwide in geosciences and social sciences. He participates in various research organizations in the field of environmental research. He is a member of the board of the Integrated Assessment Modelling Consortium (IAMC) and the Global Carbon Project. He participates in the editorial board of Climatic Change, Earth System Dynamics, and Global Environmental Change. Detlef van Vuuren has been active in various research projects and assessments. He had a coordinating role in the development of the Representative Concentration Pathways (RCPs) and Shared Socio-economic Pathways (SSPs) now used in the IPCC's assessments. Detlef van Vuuren has participated as (Coordinating) Lead Author in various assessments such as the assessment reports of IPCC, the Millennium Ecosystem Assessment, UNEP's Global Environmental Outlook, the International Assessment on Agricultural Science and Technology Development, and the OECD Environmental Outlook.

Iain Walker is a Staff Scientist at LBNL and leads the Residential Building research group. His current work focuses on developing smart ventilation technologies, indoor air quality measurement, scoring and rating, kitchen ventilation, deep energy retrofits, design and construction of zero/low energy homes, and building performance diagnostics. He is an ASHRAE Fellow and leads and contributes to many industry standards and technical committees for ASHRAE, ASTM, RESNET, and other US and international agencies.

Perry Yang is an Associate Professor and Director of Eco Urban Lab of the School of City and Regional Planning and the School of Architecture at the Georgia Institute of Technology. Perry's work focuses on promoting ecological and energy performance of cities through urban design. He has been awarded prizes in international competitions continuously from 2005 in Asian cities, including the 2009 World Games Park at Kaohsiung, Taiwan, a project opened in July 2009 and featured by CNN as an "eco-friendly" venue.

Yang published extensively on ecological urban design. A new book "https://www.elsevier.com/books/urban-systems-design/yamagata/978-0-12-816055-8" *Urban Systems Design: Creating Sustainable Smart Cities in the Internet of Things Era* that he co-edited and co-authored was published in January 2020 by Elsevier. He is the guest editor of *Environment and Planning B: Urban Analytics and City Science* for a 2019 "https://journals.sagepub.com/toc/epbb/46/8" theme issue Urban Systems Design: From Science for Design to Design in Science to explore new urban design research agenda and applications of emerging technologies, data analytics and urban automation to placemaking in the context of smart city movement.

Yang is a board member of the International Urban Planning and Environment Association (UPE), and a scientific committee member of International Conference on Applied Energy (ICAE) to chair the session on Urban Energy Systems Design for ICAE from 2014 to 2019. Prior to joining the Georgia Tech faculty, he was a Fulbright Scholar and SPURS Fellow at MIT from 1999 to 2000, and an Assistant Professor of Architecture and Urban Design at the National University of Singapore from 2001 to 2008.

Jeanne Yu is the Director of Technology Integration and the ecoDemonstrator Program for Boeing Commercial Airplanes (BCA).

Yu's responsibilities are planning and executing Product Development technology R&D portfolio to meet the needs of current and future commercial airplanes and services. This includes the innovative ecoDemonstrator Program, a cadence of flight test platforms to accelerate "learn by doing", collaboration, and technology implementation. Yu also leads Design for Environment and Operational Efficiency focused improvements for energy, emissions, materials, and community noise.

She was a key industry leader responsible for conducting first flights with sustainable biofuel blends, establishing viability of sustainable aviation biofuel for commercial aircraft. Yu has previous experience in Environmental Control Systems, air quality, 787 cabin environment, fire protection systems, and certification.

Yu has served on National Academies Transportation Research Board committees, FAA Research Engineering and Development Advisory Committee, International UK advisory for Aviation and Environment, OMEGA, and FAA Center of Excellence Aircraft — Cabin Environment. Yu was inducted as an AIAA Associate Fellow in January, 2019.

Yu received a Bachelor of Science degree in Mechanical Engineering from University of Illinois and a Master of Science in Mechanical Engineering — Thermosciences from Stanford University. Prior to joining Boeing, Yu worked at Sandia National Laboratories.

Nan Zhou is a Staff Scientist, Department Head of the International Energy Analysis Department, LBNL. In addition, Dr. Zhou is also the director of the presidential program U.S.–China Clean Energy Center-Building Energy Efficiency (CERC-BEE). She is an Advisory Board Member of Asia Pacific Energy Research Centre under APEC, as well as for APEC Sustainable Energy Center. She was appointed as the Applied Energy Editorial Board member in 2017. She was selected to serve as a Lead Author for the chapter on Mitigation and Development Pathways in the Near- to Mid-Term of the Intergovernmental Panel on Climate Change's (IPCC) Sixth Assessment Report. She received the 2017 R&D100 Award for the BEST City tool. She also received the National Excellent Young Scholar Award by Architectural Institute of Japan. Dr. Zhou's expertise includes energy and emission modeling, efficiency for buildings and appliances, and low-carbon city development. Prior to LBNL, she was an assistant professor in Japan. She has more than 220 publications including ones published in Nature Energy.

CPSIA information can be obtained
at www.ICGtesting.com
Printed in the USA
LVHW081806190422
716644LV00003B/144